ELECTRIC MACHINES

MODELING, CONDITION MONITORING, AND FAULT DIAGNOSIS

ELECTRIC MACHINES

MODELING, CONDITION MONITORING, AND FAULT DIAGNOSIS

HAMID A. TOLIYAT
SUBHASIS NANDI
SEUNGDEOG CHOI
HOMAYOUN MESHGIN-KELK

CRC Press is an imprint of the
Taylor & Francis Group, an **informa** business

CRC Press
Taylor & Francis Group
6000 Broken Sound Parkway NW, Suite 300
Boca Raton, FL 33487-2742

First issued in paperback 2017

Version Date: 20120612

ISBN 13: 978-1-138-07397-5 (pbk)
ISBN 13: 978-0-8493-7027-4 (hbk)

Library of Congress Cataloging-in-Publication Data

Electric machines : modeling, condition monitoring, and fault diagnosis / Hamid A. Toliyat ... [et al.].
p. cm.
Includes bibliographical references and index.
ISBN 978-0-8493-7027-4 (hardback)
1. Electric machinery--Reliability. 2. Machinery--Monitoring. 3. Machine parts--Failures. I. Toliyat, Hamid A.

TK2313.E44 2012
621.31'042--dc23 2012021350

Visit the Taylor & Francis Web site at
http://www.taylorandfrancis.com

and the CRC Press Web site at
http://www.crcpress.com

Contents

Preface

The development of the electric motor is one of the greatest achievements of the modern energy conversion industry. Countless electric motors are being used in our daily lives for critical service applications such as transportation, medical treatment, military operation, and communication. However, due to the fundamental limitations of material lifetime, deterioration, contamination, manufacturing defects, or damages in operations, an electrical motor will eventually go into failure mode. An unexpected failure might lead to the loss of valuable human life or a costly standstill in industry, which needs to be prevented by precisely detecting or continuously monitoring the working condition of a motor.

This book was written to provide a full review of diagnosis technologies and as an application guide for graduate and senior undergraduate students in the power electronics discipline who want to research, develop, and implement a fault diagnosis and condition monitoring scheme for better safety and improved reliability in electric motor operation. Furthermore, electrical and mechanical engineers in the industry are also encouraged to use portions of this book as a reference to understand the fundamentals of fault cause and effect and to fulfill successful implementation.

This book approaches the fault diagnosis of electrical motors through the process of theoretical analysis and then practical application. First, the analysis of the fundamentals of machine failure is presented through the winding functions method, the magnetic equivalent circuit method, and finite element analysis. Second, the implementation of fault diagnosis is reviewed with techniques such as the motor current signature analysis (MCSA) method, frequency domain method, model-based techniques, and pattern recognition scheme. In particular, the MCSA implementation method is presented in detail in the last chapters of the book, which discuss robust signal processing techniques and reference-frame-theory-based fault diagnosis implementation for hybrid vehicles as an example. These theoretical analysis and practical implementation strategies are based on many years of research and development at the Electrical Machines & Power Electronics (EMPE) Laboratory at Texas A&M University.

Hamid Toliyat
Texas A&M University
College Station, Texas

1

Introduction

Seungdeog Choi, Ph.D.
Toshiba International

The population of electric motors has greatly increased in recent years, not only in the United States but also in the world market as shown in Table 1.1 and Table 1.2. The world market is expected to be around $16.1 billion in 2011, which is assumed more than 50% growth just within 5 years [1]. Electric motors have been applied to almost every place in our daily life, such as manufacturing systems, air transportations, ground transportations, building air-conditioner systems, home energy conversion systems, various cooling systems in electrical devices, and even cell phone vibration systems.

It is also a well-known fact that the electric motors consume more than 50% of whole electrical energy demand in the United States. The annual electrical energy demand in the United States was 3,873 billion kilowatt-hours in 2008, which is expected to be further increased in every year depending on population and economic growth [11]. This data indicates that more than 1,900 billion kilowatt-hours is consumed by electric motors annually in the United States, which is the biggest energy consumption by any single electric device in modern society.

With the rapidly increased population and huge electric energy consumption, sophisticated control and reliability of motor operations from a harsh industrial environment has now been a major requirement in many industrial applications. It is especially important where an unexpected shutdown might result in the interruption of critical services such as medical, transportation, or military operations. In those applications where continuous process is needed and where down time is not tolerable, an unexpected failure of a motor might result in costly maintenance or loss of life.

As shown in Figure 1.1, the electrical motor consists of many mechanical and electrical parts, such as a rotor bar, rotor magnet, stator winding, endring, bearing, and gear box. Due to the commonly harsh industrial environments, each part of electric motors is potentially exposed to the high risk of unexpected mechanical, chemical, and electrical system failures. The reasons why electric motors fail in industry have been commonly reported as follows:

1. Post the standard lifetime
2. Wrong-rated power, voltage, and current

TABLE 1.1

Number of Motors by Application

Application	Population
Fans and pumps	3,847,161
Air compressor	632,731
Others	7,954,438
TOTAL	12,434,330

Source: US Department of Energy (2002). http://www1.eere.energy.gov/manufacturing/tech_deployment/pdfs/mtrmkt.pdf

3. Unstable supply voltage or current source
4. Overload or unbalanced load
5. Electrical stress from fast switching inverters or unstable ground
6. Residual stress from manufacturing
7. Mistakes during repairs
8. Harsh application environment (dust, water leaks, environmental vibration, chemical contamination, high temperature)

Figure 1.2 shows an example of a well known electrical motor fault such as bearing ball damage. The bearing ball is taken from the bearing module that had been diagnosed as faculty for 6 months. The main types of motor faults are commonly categorized as electrical faults, mechanical faults, and outer drive system defects, which are as follows [2–5]:

1. Electrical faults
   a. Open or short circuit in motor windings (mainly due to winding insulation failure)
   b. Wrong connection of windings
   c. High resistance contact to conductor
   d. Wrong or unstable ground

TABLE 1.2

Motor System Energy Usage by Application

Application	GWh / Yr
Fans and pumps	221,417
Air compressor	91,050
Others	262,961
TOTAL	575,428

Source: US Department of Energy (2002). http://www1.eere.energy.gov/manufacturing/tech_deployment/pdfs/mtrmkt.pdf

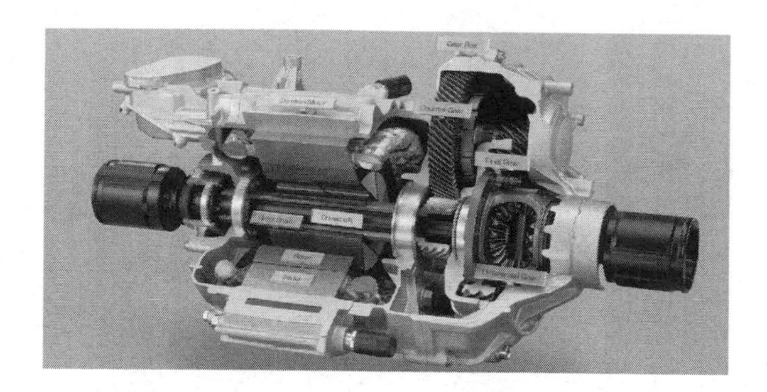

FIGURE 1.1
2009 Honda FCX Clarity Fuel Cell Vehicle test drive photo gallery. From Christine and Scott Gable, http://alternativefuels.about.com/od/fuelcellvehiclereviews/ig/ 09-Honda- FCX-Clarity-Fuel-Cell/

2. Mechanical faults
   a. Broken rotor bars
   b. Broken magnet (or partial demagnetization)
   c. Cracked end-rings
   d. Bent shaft
   e. Bolt loosening
   f. Bearing failure
   g. Gearbox failure
   h. Air-gap irregularity
3. Outer motor drive system failures
   a. Inverter system failure
   b. Unstable voltage/current source
   c. Shorted or opened supply line

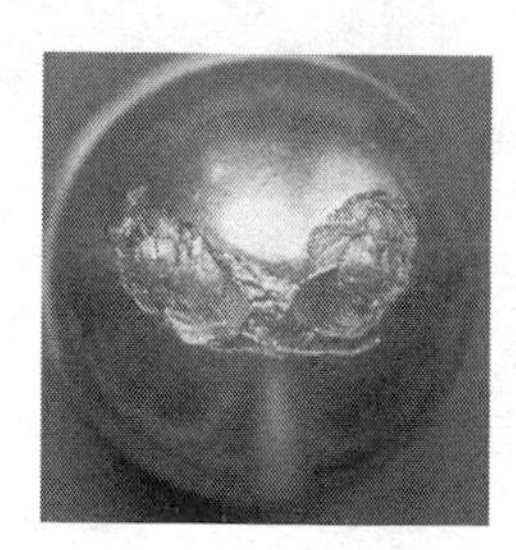

FIGURE 1.2
Bearing ball fault and subsequent fatigue damage. Vibration consultant. http://www.vibrationconsultants.co.nz/Fault%20Diagnosis.html

The bearing fault is known to make up almost 40%, stator related about 38%, rotor related about 10%, and others make up 12% of whole electrical motor fault [2–6].

The electric motor design is commonly intended to have electrical and mechanical symmetry in the stator and the rotor for better coupling and higher efficiency. Fault condition in a motor described earlier is supposed to damage the symmetrical property where fault-dependent motor operation induces an abnormal symptom during motor operation, which is described as follows [2–5]:

1. Mechanical vibration
2. Temperature increase
3. Irregular air-gap torque
4. Instantaneous output power variation
5. Acoustic noise
6. Line voltage changes
7. Line current changes
8. Speed variations

Most abnormal symptoms have been known to have specific patterns pertaining to the motor fault conditions and severity, such as particular frequency, duration, amplitude, variance, degree, and phase. Based on monitoring and analyzing the expected symptoms and their specific patterns, many motor fault diagnoses have been suggested, and there have been several commercial solutions in the industry market as shown in Figure 1.3. In particular, the vibration spectrum in Figure 1.3a is from the bearing module with defect ball in figure 1.2. Based on the spectrum monitoring technique, the bearing module is diagonosed faculty and safely removed before the system falls into catastrophic failure mode.

The various diagnosis techniques adopted in industry have been performed mainly through the following strategies [2–5].

1. Signal-based fault diagnosis
   a. Mechanical vibration analysis
   b. Shock pulse monitoring
   c. Temperature measurement
   d. Acoustic noise analysis
   e. Electromagnetic field monitoring through inserted coil
   f. Instantaneous output power variation analysis
   g. Infrared analysis
   h. Gas analysis

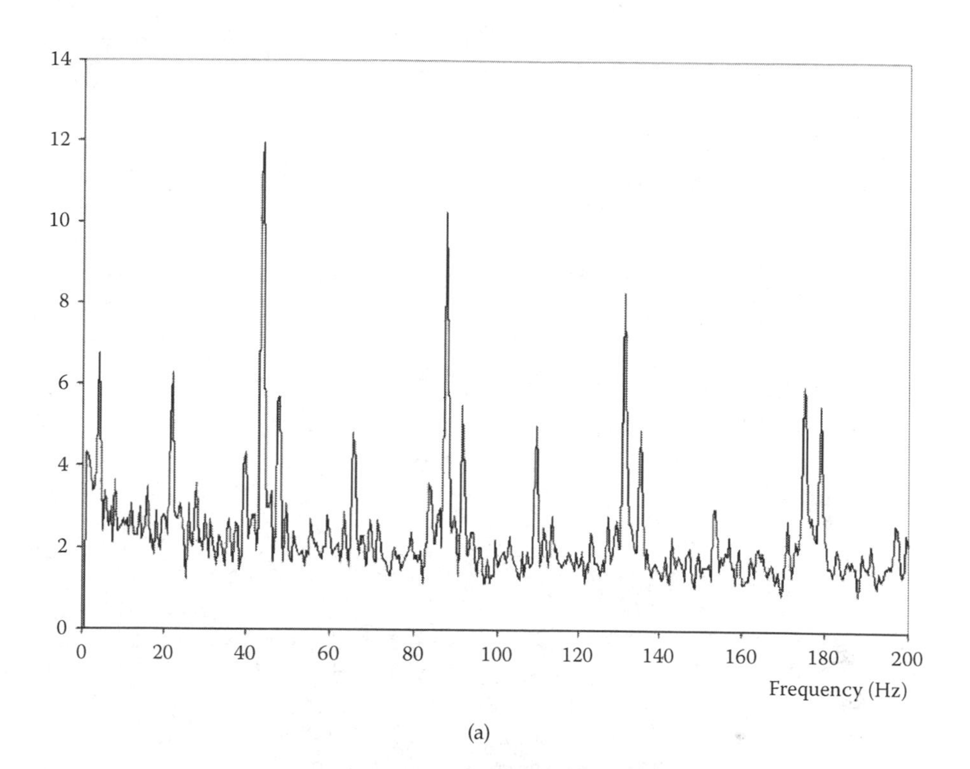

(a)

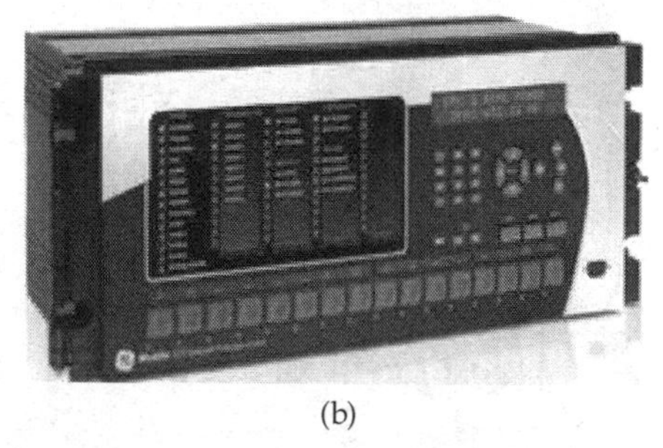

(b)

FIGURE 1.3
(a) Vibration spectrum monitoring for bearing in Figure 1.2. http://www.vibrationconsultants.co.nz/Fault%20Diagnosis.html. (b) GE motor current analysis device (from gedigitalenergy.com). http://www.gedigitalenergy.com/multilin/catalog/m60.htm

i. Oil analysis
j. Radio-frequency (RF) emission monitoring
k. Partial discharge measurement
l. Motor current signature analysis (MCSA)
m. Statistical analysis of relevant signals

2. Model-based fault diagnosis
    a. Neural network
    b. Fuzzy logic analysis
    c. Genetic algorithm
    d. Artificial intelligence
    e. Finite-element (FE) magnetic circuit equivalents
    f. Linear-circuit-theory-based mathematical models
3. Machine-theory-based fault analysis
    a. Winding function approach (WFA)
    b. Modified winding function approach (MWFA)
    c. Magnetic equivalent circuit (MEC)
4. Simulations-based fault analysis
    a. Finite-element analysis (FEA)
    b. Time-step coupled finite element state space analysis (TSCFE-SS)

The different types of fault diagnosis methods have been simultaneously applied to fine-tune the detection in industry. The fault diagnosis of electrical motors is expected to provide warning of imminent failures, diagnosing scheduling information for future preventive maintenance.

The implementation of fault diagnosis has been done with the following routine:

1. Fault detection
    a. Time-domain-based detection (mostly for power system fault diagnosis)
    b. Frequency domain-based detection (mostly for signal-based machine fault diagnosis)
    c. Accumulated data-based detection (mostly for model-based fault diagnosis)
2. Fault decision making
    a. Decide fault existence
    b. Decide fault severity
3. Feedback to motor controller or human interface
    a. Limit motor operation based on fault severity
    b. Schedule maintenance

Figure 1.4 shows the increased convergence between the energy system and modern network system in modern industry. The electrical motors in a car, ship, aircraft, building, road, or in a power system can be assumed to be

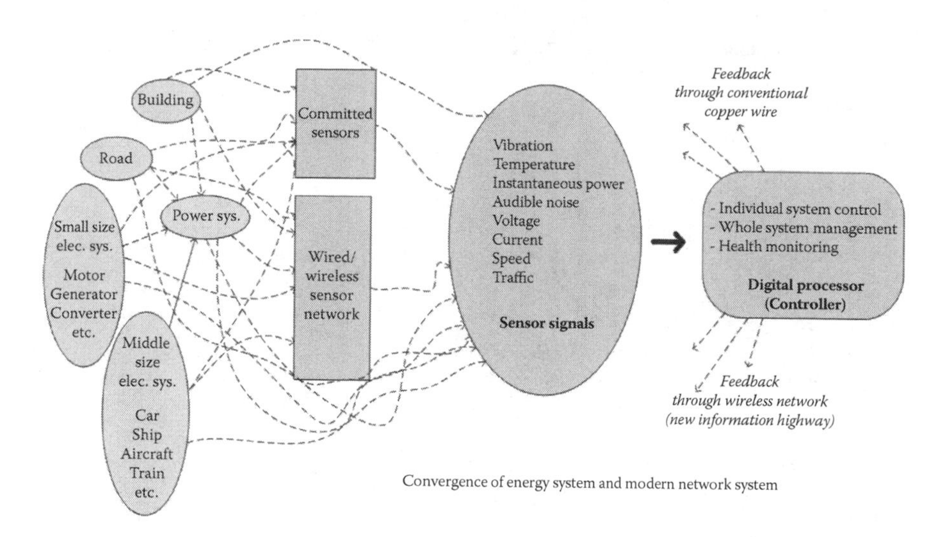

FIGURE 1.4
Convergence of energy system and modern network system. (From S. Choi, "Robust Condition Monitoring and Fault Diagnosis of Variable Speed Drive of Induction Motor," PhD dissertation, Texas A&M University, 2010. With permission.)

mostly connected to a committed sensor or wired/wireless sensor network. Those sensed signals such as vibration, current, voltage, and speed are forwarded to a close or remote microcontroller or digital processor of which the controller performs individual system control, whole system management, or health monitoring [9].

The fault diagnosis has begun to be efficiently implemented with relatively low cost by utilizing the available sensors and digital signal processor (DSP) in the wired/wireless network without extra hardware cost and with simple software implementation, which further provides the protection to middle/low power motor drive system. For example, by using the current sensor feedback, the new trend for low-cost protection applications of MCSA fault diagnosis seems to be drive-integrated fault diagnosis systems within motor drive DSP without using any external hardware [8].

This book is intended to provide fundamentals of various motor fault conditions, advanced fault modeling theory, diverse fault diagnosis techniques, and low cost DSP-based fault diagnosis implementation strategies.

The following chapters of this book are organized as follows:

- Induction of motor and synchronous motor faults in Chapter 2
- Electric motor fault modeling based on diverse theories in Chapters 3 and 4
- Various electric motor fault diagnosis techniques in Chapters 5, 6, and 7
- MCSA implementation on a microcontroller in Chapters 8, 9, and 10

References

[1] H.A. Toliyat and S.G. Campbell, *DSP-Based Electromechanical Motion Control*, Boca Raton, FL: CRC Press, 2003.

[2] G.B. Kliman, R.A. Koegl, J. Stein, R.D. Endicott, and M.W. Madden, "Noninvasive detection of broken rotor bars in operating induction motors," *IEEE Transactions on Energy Conversions*, vol. 3, pp. 873–879, December 1988.

[3] S. Nandi, H.A. Toliyat, and X. Li, "Condition monitoring and fault diagnosis of electrical machines—A review," *IEEE Transactions on Energy Conversion*, vol. 20, no. 4, pp. 719–729, December 2005.

[4] A. Siddique, G.S. Yadava, and B. Singh, "A review of stator fault monitoring techniques of induction motors," *IEEE Trans. on Energy Conversion*, vol. 20, pp. 106–114, March 2005.

[5] M. El Hachemi Benbouzid, "A review of induction motors signature analysis as a medium for faults detection," *IEEE Transactions on Industrial Electronics*, vol. 47, pp. 984–993, October 2000.

[6] Y.E. Zhongming and W.U. Bin, "A review on induction motor online fault diagnosis," *IEEE IPEMC'00*, vol. 3, pp. 1353–1358, 2000.

[7] B. Akin, U. Orguner, H. Toliyat, and M. Rayner, "Phase sensitive detection of motor fault signatures in the presence of noise," *IEEE Transactions on Industrial Electronics*, vol. 55, no. 6, June 2008.

[8] B. Akin, U. Orguner, H. Toliyat, and M. Rayner, "Low order PWM inverter harmonics contributions to the inverter fed IM fault diagnosis," *IEEE Transactions on Industrial Electronics*, vol. 55, pp. 610–619, February 2008.

[9] S. Choi, "Robust Condition Monitoring and Fault Diagnosis of Variable Speed Drive of Induction Motor," PhD dissertation, Texas A&M University, 2010.

[10] W.T. Thomson and M. Fenger, "Current signature analysis to detect induction motor faults," *IEEE Industry Applications Magazine*, vol. 7, no. 4, 2001.

[11] U.S. Energy Information Administration, "Annual Energy Outlook 2010: With Projections to 2035," Washington, DC, April 2010.

2

Faults in Induction and Synchronous Motors

Bilal Akin, Ph.D.
Texas Instruments

Mina M. Rahimian, Ph.D.
Texas A&M University

2.1 Introduction of Induction Motor Fault

This section briefly summarizes motor fault conditions and their cause, especially for the induction motor. The eccentricity related faults, broken rotor bar faults, bearing faults, and stator faults, which account for more than 90% of overall induction motor failures, are considered [1–3].

2.1.1 Bearing Faults

Bearing faults account for more than 40% of all electric motor failures [5–7]. Most of the bearings in industrial facilities run under nonideal conditions and are subject to fatigue, ambient mechanical vibration, overloading, misalignment, contamination, current fluting, corrosion, and wrong lubrication. These nonideal conditions start as marginal defects that spread and propagate on the inner raceway, outer raceways, and rolling elements (see Figure 2.1). After a while the defect becomes significant and generates mechanical vibration causing acoustic noise. Basically, bearing faults can be classified as outer raceway, inner raceway, ball defect, and cage defect, which are the main sources of machine vibration. These mechanical vibrations in the air gap due to bearing faults can be considered as slight rotor displacements, which result in instant eccentricities. Therefore, the basic fault signature frequency equation of line current due to bearing defects is adopted from eccentricity literature [10].

Mechanical vibration, infrared or thermal, and acoustic analyses are some of the commonly used predictive maintenance methods to monitor the health of the bearings to prevent motor failures.

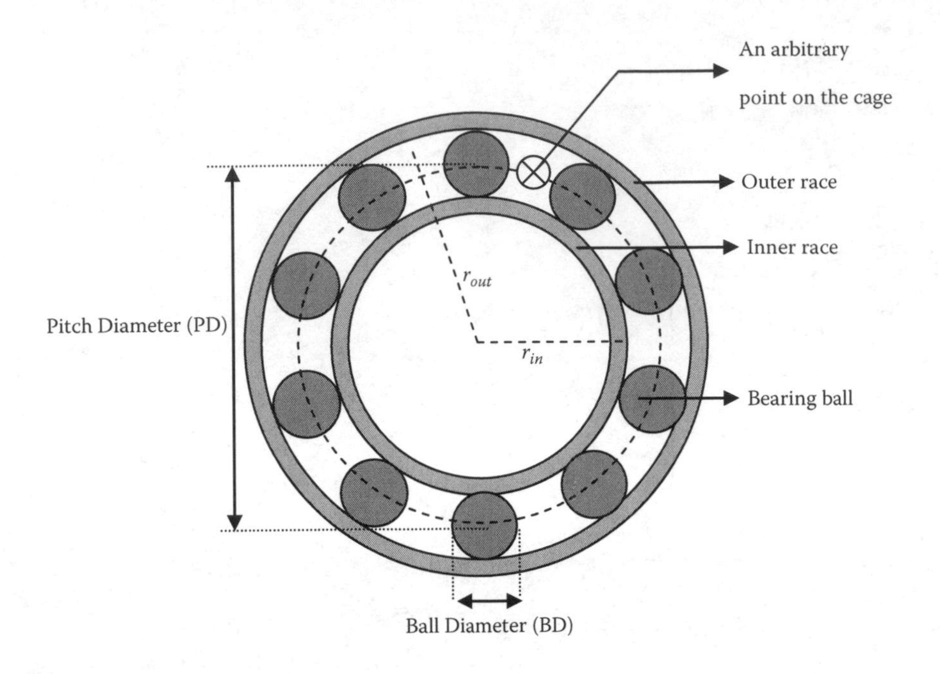

FIGURE 2.1
A typical bearing geometry.

Vibration and thermal monitoring require additional sensors or transducers to be fitted on the machines. While some large motors may already come with vibration and thermal transducers, it is not economically or physically feasible to provide the same for smaller machines. Therefore, small- to medium-size motors are checked periodically by moving portable equipment from machine to machine in all three methods. Some motors used in critical applications, such as nuclear reactor cooling pump motors, may not be easily accessible during reactor operation. The lack of continuous monitoring and accessibility are the shortcomings of the aforementioned techniques. An alternate approach based on line current monitoring has received much research attention in search of providing a practical solution to continuous monitoring and accessibility problems. Motor current monitoring provides a nonintrusive way to continuously monitor motor reliability with minimal additional cost.

Bearing faults can be classified as outer raceway, inner raceway, ball defect, and cage defect. Each fault has specific mechanical vibration frequency components that are characteristic of each defect type, which is a function of both bearing geometry and speed. The mechanical oscillations due to bearing faults change the air-gap symmetry and machine inductances like eccentricity faults. The machine inductance variations are reflected to the line current in terms of current harmonics, which are the indicators of bearing fault associated with mechanical oscillations in the air-gap.

A generic fault diagnosis tool based on discriminative energy functions is proposed by Ilonen et al. [12]. These energy functions reveal discriminative frequency-domain regions where failures are identified. Schoen [13] implemented an unsupervised, on-line system for induction motor based on motor line current. An amplitude modulation (AM) detector is developed to detect the bearing fault while it is still in an incipient stage of development in Stack et al. [14]. Ocak [15] developed a hidden Markov modeling (HMM) based bearing fault detection and fault diagnosis. Yazici and Kliman [16] proposed an adaptive statistical time-frequency method for detection of broken rotor bars and bearing faults in motors using motor line current.

2.1.2 Stator Faults

Stator faults account for 30% to 40% of all electric motor failures [2,8,9]. The stator fault can be broadly classified as the lamination or frame fault (core defect, circulation current, or ground, etc.) and the stator winding fault (winding insulation damage, displacement of conductors, etc.).

The major function of winding insulation materials normally is to withstand electric stress; however, in many cases it must also endure other stresses such as mechanical and environmental stresses [19]. In a motor, the torque is the result of the force created by current in the conductor and surrounding magnetic field. This shows that winding insulation must have electrical as well as mechanical properties to withstand mechanical stresses [20]. In addition, electromagnetic vibration at twice the power frequency, differential expansion forces due to the temperature variations following load changes, and impact forces due to electrical/mechanical asymmetries also affect the aging process [21].

Nonuniform temperature distribution in a motor will also cause mechanical destruction due to dilation. The manufacturing process itself may constitute a damaging or aging action. The electrical winding insulation must be strong enough to withstand the mechanical abuse while being wound and installed in the motor. Thus, the initial mechanical stresses are often very severe compared to the subsequent abuse the winding insulation gets in service [20].

Increased temperatures can cause a number of effects. The material may be inherently weaker at elevated temperatures and a failure may occur simply because of the melting of the material. This can be a very short time failure, because of the short length of time required for the temperature to rise to the melting point. On the other hand, long-term elevated temperature can cause internal chemical effects on material [19].

Thermal stress is probably the most recognized cause of winding insulation degradation and ultimate failure. The main sources of thermal stress in electric machinery are copper losses, eddy current, and stray load losses in the copper conductors, plus additional heating due to core losses, windage, and so forth [22]. High temperature causes a chemical reaction that makes winding insulation material brittle. Another problem is that due to sudden temperature increase, copper conductor and copper bars expand faster than winding insulation material, which causes stress on ground wall insulation [19].

Another significant effect on winding insulation aging is partial discharges (PD). Partial discharges are small electric sparks that occur within air bubbles in the winding insulation material due to nonuniform electric field distribution. Once begun, PD causes progressive deterioration of insulating materials, ultimately leading to electrical breakdown. On the other hand, motor winding insulation experiences higher voltage stresses when used with an inverter than when connected directly to the alternating current (AC) utility grid. The higher stresses are dependent on the motor cable length and are caused by the interaction of the fast rising voltage pulses of the drive and transmission line effects in the cable [23,24].

In addition to the aforementioned various causes, delaminating discharges, enwinding discharges, moisture attacks, abrasive material attacks, chemical decomposition, and radiation can also be counted as accelerating effects on aging of winding insulation [25].

Motor and generator winding insulation failures during machine operation can lead to a catastrophic machine failure resulting in a costly outage. Prevention of such an outage is a major concern for both the machine manufacturer and user, since it can result in significant loss of revenue during the outage as well as repair or replacement cost. In the literature [19,25], PD is taken as a signature of isolation aging, which begins within voids, cracks, or inclusions within a solid dielectric, at conductor–dielectric interfaces within solid or liquid dielectrics, or in bubbles within liquid dielectrics. Once begun, PD causes progressive deterioration of insulating materials, ultimately leading to electrical breakdown.

When a partial discharge occurs, the event may be detected as a very small change in the current drawn by the sample under test. PD currents are difficult to measure because of their small magnitude and short duration [25]. Therefore, PD in a motor/generator before a breakdown does not have a significant effect on the power system.

The most serious result of a major fault may not only destroy the machinery but may spread in the system and cause total failure. The most common type of fault, which is also the most dangerous one, is the breakdowns that may have several consequences. A great reduction of the line voltage over a major part of the power system will be observed. If an alternator is damaged, this might affect the whole system. For example, when a tolerable inter-turn or in-phase fault occurs, the power generation will be unbalanced and the power quality will drastically decrease. Extra harmonics will be injected to the whole system. If the alternator fault is not tolerable or it is a phase-to-phase fault, then the surge will damage the machine itself and some parts of the system. Unlike a motor connected to the utility following a few step-down transformers, the generator faults are more risky in terms of permanent damages and costly shutdowns depending on the grid structure. A motor with a tolerable inter-turn short behaves like an unbalanced load and disturbs the neighboring utility. However, an alternator failure affects the whole system, where a motor failure has limited distress on the power system. In both of these cases the power quality of the power system will be degraded.

In the literature, there are several methods for condition monitoring and protection of motors and generators. The superiority of these methods depends on the type of application, power rating of the machinery, location of the machinery, cost of machine itself and sensors, and so on [24–25].

Monitoring the temperature of the high power motor and generator stator windings, it is possible to determine if the winding is at risk of thermal deterioration. This can be done either by embedded thermocouples or thermal cameras. In addition, by monitoring the temperature, an increase in the stator temperature over time under the same operating conditions (load, ambient temperature, and voltage) can be indicative of the cooling system failure.

Ozone gas generation occurs as a consequence of PD on the stator coil. Surface partial discharges are the cause of deterioration from defective slot and end-winding stress relief coatings as well as conductive pollution. By monitoring the ozone gas concentration over time, failure mechanisms that give rise to the surface partial discharge can be detected [26]. Thus ozone monitoring does not find problems in the very early stages of deterioration. Ozone monitoring can be done periodically with inexpensive chemical detectors that are thrown away after each use. Otherwise, continuous ozone monitoring is now feasible with electronic detectors.

In addition, phase and ground fault relays are installed in a machine to prevent severe machine damage caused by winding insulation failure [20]. Another effective solution is on-line monitoring of partial discharge that warns the user before catastrophic damage. This can be done either by monitoring differential line current or using some special sensors such as antenna, high voltage capacitors on the machine terminals, or radio frequency (RF) current transformers at the machine neutral or on surge capacitor grounds. These sensors are sensitive to the high frequency signals from the PD, yet are insensitive to the power frequency voltage and its harmonics [25].

2.1.3 Broken Rotor Bar Fault

The broken rotor bar fault condition is shown in Figure 2.2, which accounts for more than 5% of all the electric motor failures in industry. Cage rotors are basically of two types: cast and fabricated. Previously, cast rotors were only used in small machines. Today, casting technology can be used even for rotors of machines in the range of thousands of kilowatts. Almost all squirrel-cage motor bars and end-rings are made of alloys of either aluminum or copper or pure copper. Copper and copper alloy rotors are usually of fabricated design. Aluminum rotors are dominantly die-cast constructions, with the bars and end-rings being cast in one machine operation. Cast rotors, although more rugged than the

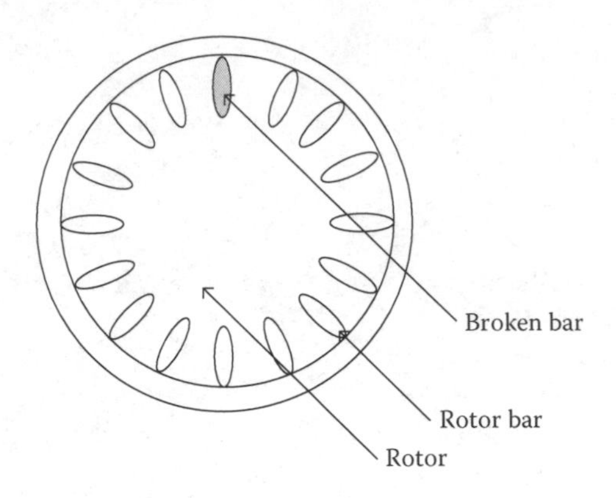

FIGURE 2.2
Broken rotor bar in an induction motor.

fabricated type, can hardly be repaired once faults like cracked or broken rotor bars develop in them.

There are a number of reasons for rotor bar and end-ring breakage. They can be caused by thermal, magnetic, dynamic, environmental, mechanical, and residual stresses. Normally, the stresses remain within the tolerance bandwidth and the motor operates properly for years. An incipient broken rotor bar condition aggravates itself almost exponentially in time as excessive current flow is expected to be concentrated on adjacent bars instead of the broken one, which provides propagated electrical stress to adjacent areas. When any of these stresses are above allowable levels, the lifetime of the motor shortens.

A broken rotor bar can be considered as rotor asymmetry [17] that causes unbalanced line currents, torque pulsation, and decreased average torque [12]. The electric and magnetic asymmetry in induction machine rotors boosts up the left sideband of supply frequency [17].

Elkasabgy et al. [29] show that broken rotor bar fault can be detected by time and frequency domain analysis of induced voltages in search coils placed in the motor. During regular operations, a symmetrical stator winding excited at frequency f_e induces rotor bar currents at sf_e frequencies [27]. When an asymmetry is introduced in the rotor structure, the backward rotating negative sequence $-sf_e$ component starts the chain electrical and mechanical interactions between the rotor and stator of the motor. Initially, stator electromotive force (EMF) at frequency $(1 - 2s)f_e$ is induced that causes torque and speed ripples. Afterward, torque and speed ripples are reflected to the stator as line current oscillations at frequency $(1 + 2s)f_e$. Next, $(1 + 2s)f_e$ component induces rotor currents at $\pm 3sf_e$ and this chain reaction goes on until completely being filtered by the rotor inertia. A parameter-estimation-based broken rotor bar detection is reported in [30]. The harmonics at the stator terminal voltages immediately after switching off the motor can be used as a diagnostic method [31].

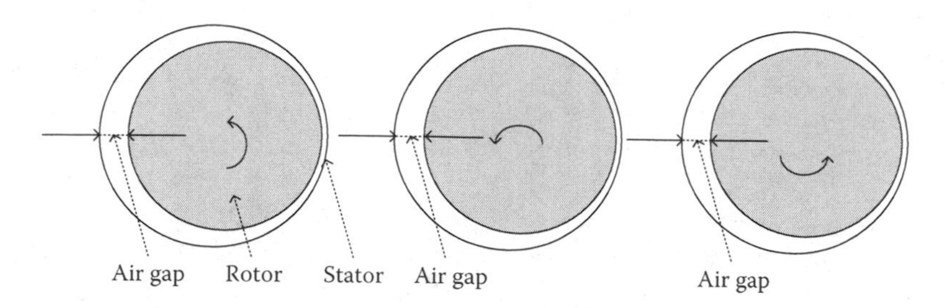

FIGURE 2.3
Static eccentricity.

2.1.4 Eccentricity Fault

Air-gap eccentricity is known as a condition that occurs when there is a non-uniform distance between the rotor and stator in the air-gap. When there is an eccentricity in the air-gap, varying inductances cause unbalanced magnetic flux within the air-gap that creates fault harmonics in the line current, which can be identified in the spectrum. There are two types of eccentricity faults: static eccentricity and dynamic eccentricity as shown in Figure 2.3 and Figure 2.4. When static eccentricity occurs, the centerline of the shaft is at a constant offset from the center of the stator or the rotor is misaligned along the stator bore. On the other hand when dynamic eccentricity occurs, the centerline of the shaft is at a variable offset from the center of the stator or minimum air-gap revolves with the rotor. If the distance between the stator bore and rotor is not equal throughout the entire machine, varying magnetic flux within the air-gap creates imbalances in the current flow, which can be identified in the current spectrum. Improper mounting, a loose or missing bolt, misalignment, or rotor unbalance might be causes of air-gap eccentricity.

Eccentricity is a quite well-known problem and analytical results supported by experiments have already been reported. In the literature, there are several successful works reporting fault diagnosis of eccentricity based on line current measurement. Unlike bearing faults, it is easier to diagnose eccentricity even for inverter-fed machine cases due to their high amplitude

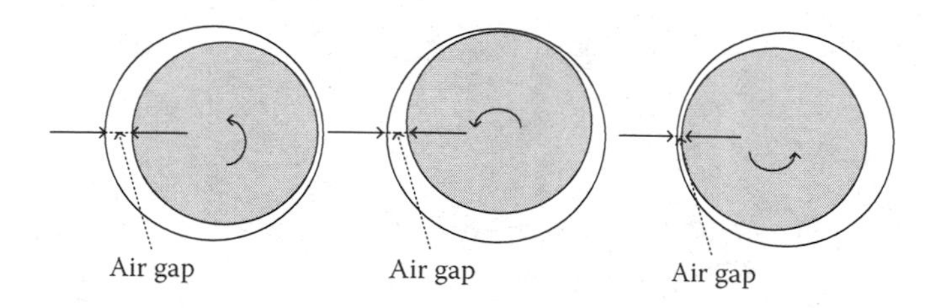

FIGURE 2.4
Dynamic eccentricity.

of fault signatures with respect to the noise floor in the line current spectrum. Because both static and dynamic eccentricities tend to coexist in practice, only mixed eccentricity is considered to show the effects of inverter harmonics. Magnetic field in the air-gap of an eccentric motor is always nonuniform. Since the flux linkages in the air-gap oscillate with synchronous frequency, any additional harmonics oscillating at the speed due to nonuniform structure are expected to take place at rotating frequency sidebands of the synchronous frequency.

2.2 Introduction of Synchronous Motor Fault Diagnosis

This section summarizes the important electrical and mechanical failures in the synchronous machines and the corresponding diagnosis techniques proposed in the literature. Some failures, like stator inter-turn faults, bearing faults, and eccentricities, are common in all types of synchronous machines. However, some faults like rotor winding faults, broken damper bars, or end-rings, are specific to wound rotor synchronous machines, and demagnetization faults are limited to permanent magnet synchronous machines (PMSMs).

Like induction machines, synchronous machines are subject to many different types of mechanical and electrical faults, which can broadly be classified into the following: (1) open or short circuit in one or more turns of a stator winding; (2) open or short circuited rotor winding in wound rotor synchronous machines; (3) broken damper bars or end-rings; (4) eccentricities; (5) rotor mechanical faults such as bearing damage, bent shaft, and misalignment; and (6) demagnetization fault in PMSMs.

Each of these fault conditions produces specific symptoms during motor operation, which can be described as follows:

1. Unbalanced line currents and air-gap voltages
2. Excessive temperature
3. Audible noise and motor mechanical vibration
4. Lower average torque
5. Higher torque pulsations
6. Increased losses

Some of the fault conditions in synchronous machines have similar causes and symptoms of the same faults in the induction machines, which have been discussed in Section 2.1.

2.2.1 Damper Winding Fault

To produce torque in synchronous machines, the rotor must be turning at synchronous speed, which is the speed of the stator field. At any other speed, the rotating field of stator poles will not be synchronized with rotor poles, but first attracts, and then repels them. This condition produces no average torque and the machine will not start. Using a direct current (DC) motor or a damper winding, the machine can be brought near to the synchronous speed. Damper windings, as shown in Figure 2.5, consist of heavy copper bars, with the two ends shorted together, installed in rotor slots. The currents induced in the bars interact by the rotating air-gap field and produces torque. In other words, the machine is started as an induction motor [33]. The field winding is excited by a direct current when the machine is brought up to the synchronous speed. When the load is suddenly changed, an oscillatory motion will be superimposed on the normal synchronous rotation of the shaft. The damper winding helps damp out these oscillations.

Diagnostics of broken damper bars in synchronous machines has not been covered as widely as the other faults like eccentricity and inter-turn faults [34–37]. During transience, the electromagnetic behavior of asynchronous machines with damper winding is similar to that of an induction machine. During transient time, when the machine accelerates from zero speed to synchronous speed, a significant current flows in the damper winding. Excessive

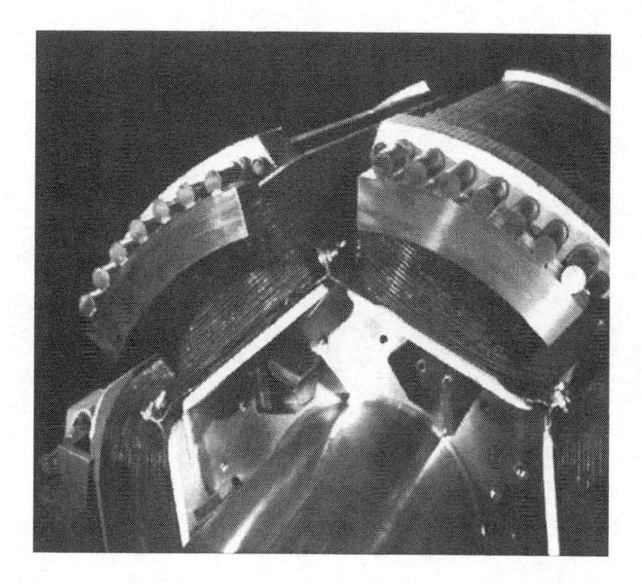

FIGURE 2.5
A salient pole synchronous machine with damper winding and interrupted end-ring. (Courtesy of TECO-Westinghouse.)

start–stop cycles or frequent load or speed changes can cause the breakage of the damper bars.

An on-line fault diagnosis method for detection of broken rotor bars has been proposed by Kramer [38] using flux probe and finite element (FE) modeling. For the squirrel-cage induction machines, several methods of detection of broken rotor bars have been reported in the literature. It has been found that in squirrel-cage induction machines when a bar breaks some of the current that would have flowed in that bar will flow into the two adjacent bars on either side. This could result in breakage of several bars [39]. Similar effects have been reported for the converter-fed synchronous machines with broken damper bars [40]. Winding function analysis and time-stepping FE analysis have been used to study the broken damper bars and end-rings [38,41].

For detecting the broken damper bars, flux probes can be attached to the stator bore surface for measurement of air-gap flux waveform during acceleration from standstill to rated speed [38]. Another method is the separation of pole voltages of the field winding according to its polarity. This way, the difference of the pole voltages can be determined. The main field of a symmetrical built machine disappears in the difference voltage, but the difference voltage, which is caused by perturbed field of the missing damper bar, will remain [42].

2.2.2 Demagnetization Fault in Permanent Magnet Synchronous Machines (PMSMs)

In comparison to other types of alternating current (AC) motors, permanent magnet synchronous Machines (PMSMs) are becoming more popular in applications with high-speed operation and precise torque control. The demagnetization phenomenon is mainly due to armature reaction, especially in high torque conditions. Some other advantageous features include high efficiency, low noise, high torque to current ratio, high power to weight ratio, and robustness.

During the normal operation of the PMSM, the inverse magnetic field produced by the stator current opposes the permanent magnets' remanent induction. When this phenomenon is repeated, the permanent magnets will be demagnetized. This demagnetization can be all over the pole (complete demagnetization), or on a part of the pole (partial demagnetization). High temperature can also demagnetize the magnet. Stator winding short-circuit fault may partially demagnetize a surface mount magnet. Partial demagnetization causes magnetic force harmonics, noise, and mechanical vibration, causing unbalanced magnetic pull in the machine.

The demagnetization effects on the parameters of the motor, such as cogging torque, torque ripple, back-EMF, and load angle curve, were investigated by Ruiz et al.[43]. For steady-state analysis under demagnetization conditions, fast Fourier transform (FFT) of the stator current is used for frequency analysis. Time-frequency analysis methods have been used for nonstationary conditions. These techniques, such as short-time Fourier transform (STFT), continuous wavelet transform (CWT), and discrete wavelet transform (DWT), require the proper selection of the parameters such as window size and coefficients.

Field reconstruction method (FRM) can also be used to detect the demagnetization fault in PMSMs. The flux linkages of the stator phases, which are calculated by FRM, are used to monitor the faults [44].

2.2.3 Eccentricity Fault

The fundamentals of the eccentricity fault in a synchronous motor are the same as that of the induction motor. Air-gap eccentricity arises when there is a nonuniform distance between the stator and the rotor. The nonuniform air-gap causes the varying inductances giving rise to unbalanced magnetic flux within the air-gap. This creates fault harmonics in the line current, which can be identified in the frequency spectrum. When eccentricity becomes significant, the resulting unbalanced radial forces can cause stator to rotor rub, and this can result in damage of the stator and rotor.

There are two types of eccentricity in the synchronous motor as in the induction motor: static and dynamic. In the case of static eccentricity, the centerline of the shaft is at a constant offset from the center of the stator. Therefore, the nonuniform air-gap does not vary in time. On the other hand, when dynamic eccentricity occurs, the centerline of the shaft is at a variable offset from the center of the stator and the air-gap length changes as the rotor rotates dynamically. In reality, both static and dynamic eccentricities tend to coexist. Improper mounting, the noncircularity of the stator core, a loose or missing bolt, a bent rotor shaft or misalignment, bearing wear, and rotor unbalance might be causes of air-gap eccentricity.

Various fault diagnosis methods for eccentricity fault detection in synchronous machines have been proposed in literature. The modified winding function approach (MWFA) accounting for all space harmonics, and the FE method have been used to model the salient pole synchronous machines. These models show the effect of dynamic air-gap eccentricity on the performance of a salient pole synchronous machine [45].

Ebrahimi et al. present a method of detecting static eccentricity (SE), dynamic eccentricity (DE), and mixed eccentricity (ME) in three-phase PMSMs [46]. The nominated index is the amplitude of sideband components with a particular frequency pattern in the stator current spectrum. The occurrence, as well as the type and percentage, of eccentricity can be determined using this index. After determination of the correlation between the index and the SE and DE, the type of the eccentricity is determined by a *k*-nearest neighbor classifier. Then a three-layer artificial neural network is employed to estimate the eccentricity degree and its type.

Le Roux et al. investigate implementation and detection of rotor faults such as static and dynamic eccentricities and broken magnets in permanent magnet synchronous machines [47]. A new flux estimation method is developed that does not require the measurement of the rotor position or speed. Stator currents and voltages are used for detection of these and other rotor faults.

2.2.4 Stator Inter-Turn Fault

One of the most common failures in synchronous motors is the inter-turn short circuit in one of the stator coils. Stator failures are essentially due to electrical, mechanical, thermal, and environmental stresses acting on the stator. The most recognized cause of winding insulation degradation and ultimate failure is thermal stress. The dielectric, corona, tracking, and transient voltage conditions are some of the electrical stresses leading to inter-turn short circuit failures [48].

In the case of an inter-turn fault in stator winding, the symmetry of the machine is destroyed. This produces a reverse rotating field that decreases the output torque and increases losses per ampere of fundamental frequency of the positive sequence current. The stator faults in synchronous reluctance motors (SynRM) under steady-state operating conditions have been studied in [49]. The detailed modeling of the faulted machine has been carried out using a modified winding function approach (MWFA). Monitoring the stator current in the presence of such faults shows that odd triple harmonics are increased in the line current of SynRM with inter-turn fault. The line current of the faulty phase increases further when the number of shorted turns goes up. The increase of the 9th harmonic seems to be a good indication of the inter-turn fault.

Stator inter-turn faults in a salient pole synchronous motor can be detected by analyzing the field current of the machine. Some of the even harmonics in the field current have been reported to increase with stator inter-turn faults. Due to internal structural asymmetries of the field winding, some of these components clearly increased with stator inter-turn fault. The findings are helpful to detect faults involving few turns without ambiguity, in spite of supply unbalance and time harmonics [35].

For analyzing internal phase and ground faults in stator winding, a mathematical model for a synchronous machine has been presented by Reichmeider et al. [50]. This method employs a direct phase representation, using a traditional coupled circuit approach.

A specific frequency pattern of the stator current is derived for short-circuit fault detection in PMSMs [51]. The amplitude of the side-band components at these frequencies is used to determine the number of short-circuited turns. Using the mutual information index, the relation between the nominated criterion and the number of short-circuited turns is specified. The occurrence and the number of short-circuited turns are predicted using support vector machine (SVM) as a classifier.

The two-reaction theory is well suited for computer modeling synchronous machines. However, in the derivation of the *dq*0 model of a synchronous machine, the machine windings are assumed to be sinusoidally distributed. This implies that all higher space harmonics produced in the case of an internal fault, the stator windings no longer have the characteristics of sinusoidally distributed windings. The faulted windings will produce stronger space harmonics. Moreover, the symmetry between the machine windings will no longer be present. Therefore, the conventional *dq*0 model is not suited to analyze internal faults.

Inter-turn faults of a synchronous machine can be modeled based on the actual winding arrangements. This method, which is known as winding function approach, calculates the machine inductances directly from the machine winding distribution. Using this model, the space harmonics produced by the machine windings are taken into account [52]. Abdallah et al. use a winding function approach to simulate inter-turn faults in stator windings of the permanent magnet synchronous machines [53].

The time-stepping finite element method (FEM) is another analysis method to study a synchronous machine with inter-turn fault. Vaseghi et al. employ FEM for internal fault analysis of a surface-mounted permanent-magnet synchronous machine [54]. It is used for magnetic field study and determining the machine parameters under various fault conditions and the effect of machine pole number and number of faulted turns on machine parameters.

2.2.5 Rotor Inter-Turn Fault

Rotor winding inter-turn fault is a common electrical fault in synchronous machines. Its existence may result in serious problems such as high rotor current, high winding temperature, low reactive output power, distorted voltage waveform, and mechanical vibration. The rotor winding inter-turn fault is mainly caused by poor manufacturing or operating conditions such as loose rotor end winding, loose spacer block, poor trimming of soldered joint, deformation of high-speed rotor winding due to centrifugal force, overheating, and poor insulation.

There are many studies about fault diagnosis of rotor winding turn-to-turn faults. One method is based on indirect measurement of the impedance of the rotor field winding during operation [55]. This method is useful when the number of shorted turns is significant.

Some methods are based on detection of flux asymmetry created by shorted turns by applying alternating current to the field [56]. This method is accurate but not easy to implement because it requires removing the rotor from the bore.

Some reliable methods based on direct measurement of the air-gap magnetic flux can be applied to the machine in operation [57]. The flux is measured by a search coil installed in the air gap.

The neural network models of machines can be used to detect the rotor turn faults. This method requires training data through simulation or experiment. A mathematical model of the machine is needed for simulated data. The experimental training data can be acquired using a machine in which the rotor turns can be shorted. Streifel et al. propose a method based on traveling wave [58]. This method along with neural network feature extraction and novelty detection algorithm is used for fault diagnosis of short-circuited windings in any rotating machinery and other equipment containing symmetrical windings.

The terminal parameters are affected by the fault condition of the rotor winding, but it is difficult to relate them together by accurate mathematical expressions. An artificial neural network method is investigated by Hongzhong et al. for rotor shorted winding fault diagnosis [59]. Since it is difficult to find the faulty samples in practical applications, these samples are gained through calculation. Using this method, the severity of the fault can be detected, but the location of the fault cannot be determined.

2.2.6 Bearing Fault

Even under normal operating conditions with balanced load and good alignment, bearing failures may take place. Flaking of bearings might occur when fatigue causes small pieces to separate from the bearing. Sometimes bearing faults are considered as rotor asymmetry faults, which are usually covered under the eccentricity-related faults. The bearing failures have been reported frequently in industry. Different techniques for a joint time frequency analysis and an experimental study of detection and fault diagnosis of damaged bearings on a PMSM were investigated by Rosero et al. [60]. When the motor is running under nonstationary conditions, conventional signal processing methods such as FFT in motor current signature analysis (MCSA) do not work well. In such conditions, the stator current can be analyzed by means of STFT and Gabor spectrogram for detecting the bearing damage.

Another fault diagnosis method for detecting bearing fault in PMSM based on frequency response analysis is proposed by Pacas et al. [61]. The torque and velocity signals of the machine will be periodically disturbed when the bearing is damaged. These disturbances cause the frequency response of the mechanical system to change at specific frequencies. Utilizing the velocity of the motor and the torque-generating component of the stator current (i_q), the frequency response of the machine in the closed loop speed control can be derived. The frequency response analysis proposed in this study yields more reliable fault detection results than the FFT analysis.

References

[1] S. Nandi, H.A. Toliyat, and X. Li, "Condition monitoring and fault diagnosis of electrical machines—A review," *IEEE Transactions on Energy Conversion*, vol. 20, no. 4, pp. 719–729, December 2005.

[2] A. Siddique, G.S. Yadava, and B. Singh, "A review of stator fault monitoring techniques of induction motors," *IEEE Transactions on Energy Conversion*, vol. 20, pp. 106–114, March 2005.

[3] M. El Hachemi Benbouzid, "A review of induction motors signature analysis as a medium for faults detection," *IEEE Transactions on Industrial Electronics*, vol. 47, pp. 984–993, October 2000.

[4] Y.E. Zhongming and W.U. Bin, "A review on induction motor online fault diagnosis," *IEEE IPEMC '00*, vol. 3, pp. 1353–1358, 2000.

[5] W. Zhou, T.G. Habetler, and R.G. Harley, "Incipient bearing fault detection via motor stator current noise cancellation using Wiener Filter," *IEEE Transactions on Industrial Application*, vol. 45, pp. 1309–1317, July/August 2009.

[6] W. Zhou, T.G. Habetler, and R.G. Harley, "Bearing fault detection via stator current noise cancellation and statistical control," *IEEE Transactions on Industrial Electronics*, vol. 55, no. 12, pp. 4260–4469, December 2008.

[7] J. Liu, W. Wang, and F. Golnaraghi, "An extended wavelet spectrum for bearing fault diagnostics," *IEEE Transactions on Instrumentation and Measurement*, vol. 57, pp. 2801–2812, December 2008.

[8] S.M.A. Cruz, H.A. Toliyat, and A.J.M. Cardoso, "DSP implementation of the multiple reference frames theory for the diagnosis of stator faults in a DTC induction motor drive," *IEEE Transactions on Energy Conversion*, vol. 20, no. 2, pp. 329–335, June 2005.

[9] D. Shah, S. Nandi, and P. Neti, "Stator inter-turn fault detection of doubly-fed induction generators using rotor current and search coil voltage signature analysis," *IEEE Transactions on Industry Applications*, vol. 45, no. 5, pp. 1831–1842, September–October 2009.

[10] R. Schoen, T. Habetler, F. Kamran, and R. Bartfield, "Motor bearing damage detection using stator current monitoring," *IEEE Transactions on Industry Applications*, vol. 31, no. 6, pp. 1274–1279, November/December 1995.

[11] L. Eren, "Bearing Damage Detection via Wavelet Package Decomposition of Stator Current," PhD dissertation, School of Electrical Engineering, University of Missouri-Columbia, 2002.

[12] J. Ilonen, J.-K. Kamarainen, T. Lindh, J. Ahola, H. Kälviäinen, and J. Partanen, "Diagnosis tool for motor condition monitoring," *IEEE Transactions on Industry Applications*, vol. 41, pp. 963–971, April 2005.

[13] R.R. Schoen, B.K. Lin, F.G. Habetter, H.J. Shlog, and S. Farag, "An unsupervised on-line system for induction motor fault detection using stator current monitoring," *IEEE Transactions on Industry Applications*, vol. 31, no. 6, pp. 1280–1286, November/December 2005.

[14] J.R. Stack, R.G. Harley, and T.G. Habetler, "An amplitude modulation detector for fault diagnosis in rolling element bearings," *IEEE Transactions on Industrial Electronics*, vol. 51, no. 5, pp. 1097–1102, October 2005.

[15] H. Ocak, "Fault Detection, Diagnosis and Prognosis of Rolling Element Bearings: Frequency Domain Methods and Hidden Markov Modeling," PhD dissertation, School of Electrical Engineering, Case Western Reserve University, Cleveland, OH, 2004.

[16] B. Yazici and G.B. Kliman, "An adaptive statistical time-frequency method for detection of broken bars and bearing faults in motors using stator current," *IEEE Transactions on Industry Applications*, vol. 35, no. 2, pp. 442–452, March 1999.

[17] S. Nandi, H. Toliyat, and X. Li, "Condition monitoring and fault diagnosis of electrical motors—A review," *IEEE Transactions on Energy Conversions*, vol. 20, no. 4, pp. 719–729, December 2005.

[18] A.V. Oppenheim and R.W. Schafer, *Discrete-Time Signal Processing*, New York: Prentice-Hall, 1989.

[19] G. Stone and J. Kapler, "Stator winding monitoring," *IEEE Industry Applications Magazine*, September/October 1998.

[20] E.L. Brancato, "Insulation aging," *IEEE Transactions on Dielectrics and Electrical Insulation*, vol. EI-13, no. 4, August 1978.

[21] *Handbook to Assess Stator Insulation Mechanism*, EPRI EL-5036, vol. 16, New York: Power Plant Electrical Reference Series.

[22] J. Douglas, "Hydro generator failure," *IEEE Power Engineering Review*, vol. 8, no. 11, pp. 4–6, November 1988.

[23] G. Stone, S. Campbell, and S. Tetreault, "Inverter-fed drives: Which motor stators are at risk," *IEEE Industry Application Magazine*, vol. 6, pp. 17–22, September/October 2000.

[24] B. Fernando, "Online stator winding fault diagnosis in inverter-fed AC machines using high-frequency signal injection," *IEEE Transactions on Industry Application*, vol. 39, no. 4, pp. 1109–1117, August 2003.

[25] S. Lee and G.B. Kliman, "An online technique for monitoring the insulation condition of ac machine stator windings," *IEEE Transactions on Energy Conversions*, vol. 20, no. 4, pp. 737–745, December 2005.

[26] G.C. Stone, "Advancements during the past quarter century in on-line monitoring of motor and generator winding insulation," *IEEE Transactions on Dielectrics and Electrical Insulation*, vol. 9, no. 5, pp. 746–751, October 2002.

[27] F. Filippetti, G. Franceschini, C. Tassoni, and P. Vas, "AI techniques in induction machines diagnosis including the speed ripple effect," *IEEE Transactions on Industry Applications*, vol. 34, pp. 98–108, 1998.

[28] G.B. Kliman, R.A. Koegl, J. Stein, R.D. Endicott, and M.W. Madden, "Noninvasive detection of broken rotor bars in operating induction motors," *IEEE Transactions on Energy Conversions*, vol. 3, pp. 873–879, December 1988.

[29] N.M. Elkasabgy, A.R. Eastham, and G.E. Dawson, "Detection of broken bars in the cage rotor on an induction machine," *IEEE Transactions on Industry Applications*, vol. 28, no. 1, pp. 165–171, January/February 1992.

[30] K.R. Cho, J.H. Lang, and S.D. Umans, "Detection of broken rotor bars in induction motors using state and parameter estimation," *IEEE Transactions on Industry Applications*, vol. 28, no. 3, pp. 702–709, May/June 1992.

[31] J. Milimonfared, H.M. Kelk, S. Nandi, A.D. Minassians, and H.A. Toliyat, "A novel approach for broken-rotor-bar detection in cage induction motors," *IEEE Transactions on Industry Applications*, vol. 35, no. 5, pp. 1000–1006, September/October 1999.

[32] D.G. Dorrell, W.T. Thomson, and S. Roach, "Analysis of airgap flux, current, vibration signals as a function of the combination of static and dynamic air gap eccentricity in 3-phase induction motors," *IEEE Transactions on Industry Applications*, vol. 33, pp. 24–34, January 1997.
[33] G. McPherson and R.D. Laramore, *Electrical Machines and Transformers*, 2nd ed., New York: John Wiley, 1990.
[34] P. Neti and S. Nandi, "Analysis and modeling of a synchronous machine with structural asymmetries," *IEEE-Canadian Conference on Electrical and Computer Engineering*, Ottawa, Canada, May 2006.
[35] P. Neti and S. Nandi, "Stator inter-turn fault detection of synchronous machines using field current signature analysis," *IEEE Industry Applications Conference*, Tampa, FL, October 2006.
[36] N.A. Al-Nuaim and H.A. Toliyat, "A method for dynamic simulation and detection of dynamic air gap eccentricity in synchronous machines," *IEEE International Electric Machines and Drives Conference*, May 1997.
[37] H.A. Toliyat and N.A. Al-Nuaim, "Simulation and detection of dynamic air-gap eccentricity in salient-pole synchronous machines," *IEEE Transactions Industry Applications*, vol. 35, no. 1, pp. 86–93, January–February 1999.
[38] H.C. Kramer, "Broken damper bar detection studies using flux probe measurements and time-stepping finite element analysis for salient pole synchronous machines," *IEEE International Symposium on Diagnostics for Electric Machines, Power Electronics and Drives, SDEMPED* '03, pp. 193–197, August 2003.
[39] G.B. Kliman, and R.A Koegl, "Noninvasive detection of broken rotor bars in operating induction motors," *IEEE Transactions Energy Conversion*, vol. 3, no. 4, pp. 873–879, December 1988.
[40] S.B. Jovanovski, "Calculation and testing of damper-winding current distribution in a synchronous machine with salient poles," *IEEE Transactions on Power Apparatus and Systems*, vol. PAS-88, no. 11, pp. 1611–1619, November 1969.
[41] M.M. Rahimian and K. Butler-Purry, "Modeling of synchronous machines with damper windings for condition monitoring," *Electric Machines and Drives Conference, IEMDC* 2009.
[42] J.P. Bacher, "Detection of broken damper bars of a turbo generator by the field winding," *International Conference on Renewable Energy and Power Quality, ICREPC* 2004.
[43] J.R.R. Ruiz, J.A. Rosero, A.G. Espinosa, and L. Romeral, "Detection of demagnetization faults in permanent-magnet synchronous motors under nonstationary conditions," *IEEE Transactions on Magnetics*, vol. 45, no. 7, pp. 2961–2969, July 2009.
[44] A. Khoobroo and B. Fahimi, "A novel method for permanent magnet demagnetization fault detection and treatment in permanent magnet synchronous machines," *Applied Power Electronics Conference and Exposition*, pp. 2231–2237, 2010.
[45] H.A. Toliyat and N.A. Al-Nuaim, "Simulation and detection of dynamic airgap eccentricity in salient-pole synchronous machines," *IEEE Transactions on Industry Applications*, vol. 35, no 1, pp.86–93, January/February 1999.
[46] B.M. Ebrahimi, J. Faiz, and M.J. Roshtkhari, "Static-, dynamic-, and mixed-eccentricity fault diagnoses in permanent-magnet synchronous motors," *IEEE Transactions on Industrial Electronics*, vol. 56, no. 11, pp. 4727–4739.

[47] W. le Roux, R.G. Harley, and T.G. Habetler, "Detecting rotor faults in low power permanent magnet synchronous machines," *IEEE Transactions on Power Electronics*, vol. 22, no. 1, pp. 322–328, January 2007.
[48] A. Siddique, G.S. Yadava, and B. Singh, "A review of stator fault monitoring techniques of induction motors," *IEEE Transactions On Energy Conversion*, vol. 20, pp. 106–114, March 2005.
[49] P. Neti and S. Nandi, "Stator inter-turn fault analysis of reluctance synchronous motor," *Canadian Conference on Electrical and Computer Engineering*, pp. 1283–1286, 2005.
[50] P.P. Reichmeider, C.A. Gross, D. Querrey, D. Novosel, and S. Salon, "Internal faults in synchronous machines, Part I: The machine model," *IEEE Transactions on Energy Conversion*, vol. 15, no. 4, pp. 376–379, 2000.
[51] B. M. Ebrahimi and J. Faiz, "Feature extraction for short circuit fault detection in permanent magnet synchronous motors using stator current monitoring," *IEEE Transactions on Power Electronics*, vol. 25, no. 10, pp. 2673–2682, 2010.
[52] T. Xiaoping, L.A. Dessaint, M.E. Kahel, and A.O. Barry, "A new model of synchronous machine internal faults based on winding distribution," *IEEE Transactions on Industrial Electronics*, vol. 53, no. 6, pp. 1818–1828, 2006.
[53] A. Ali Abdallah, J. Regnier, J. Faucher, and B. Dagues; "simulation of internal faults in permanent magnet synchronous machines," *International Conference on Power Electronics and Drive Systems*, vol. 2, pp. 1390–1395, 2005.
[54] B. Vaseghi, B. Nahid-Mobarakeh, N. Takorabet, and F. Meibody-Tabar, "Modeling of non-salient pm synchronous machines under stator winding inter-turn fault condition: Dynamic model–FEM model," *Vehicle Power and Propulsion Conference*, pp. 635–640, 2007.
[55] *Rotating Machinery Insulation Test Guide*, Doble Engineering Company, 1985.
[56] *Handbook to Assess the Insulation Condition of Large Rotating Machines*, EPRI Power Plant Electrical Reference Series, vol. 16, pp. 5-24–5-31.
[57] D. R. Albright, "Interturn short-circuit detection for turbine generator rotor winding," *IEEE Transactions on Power and Apparatus Systems*, vol. PAS-90, pp. 478–483, 1971.
[58] R.J. Streifel, R.J. Marks, M.A. El-Sharkawi, and I. Kerszenbaum, "Detection of shorted turns in the field winding of turbine generator rotors using novelty detector development and field test," *IEEE Transactions on Energy Conversion*, vol. 11, no. 2, pp. 312–317, 1996.
[59] M. Hongzhong, D. Yuanyuan, P. Ju, and Z. Limin, "The application of ANN in fault diagnosis for generator rotor winding turn-to-turn faults," IEEE Power and Energy Society General Meeting—Conversion of Electrical Energy in the 21st Century, 2008.
[60] J. Rosero, J. Cusido, A. Garcia Espinosa, J.A. Ortega, L. Romeral, "Broken bearings fault detection for a permanent magnet synchronous motor under non-constant working conditions by means of a joint time frequency analysis," IEEE International Symposium on Industrial Electronics, 2007.
[61] M. Pacas, S. Villwock, and R. Dietrich, "Bearing damage detection in permanent magnet synchronous machines," *IEEE Energy Conversion Congress and Exposition*, pp. 1098–1103, 2009.

3

Modeling of Electric Machines Using Winding and Modified Winding Function Approaches

Subhasis Nandi, Ph.D.
University of Victoria

3.1 Introduction

A rotating electric machine typically consists of a hollow cylindrical static structure (called stator in alternating current machines and field in direct current machines) and a rotating cylinder (called rotor in alternating current machines and armature in direct current machines) mounted on bearings and placed inside the hollow of the static structure. Both the static and the rotating members are made of laminated steel and they carry current carrying copper or aluminum conductors for the production of torque or voltage and conversion of electric energy to mechanical energy and vice versa.

Since an electric machine is an electromagnetic device, the best possible way to analyze it is to obtain the electromagnetic field distribution of the machine. This requires solution of Laplace's or Poisson's equation, which even for the best computer available today is an onerous task given the complicated structure of even the simplest machine.

Electric fault diagnosis often requires analysis of harmonics in machine line current, flux, torque, and speed. Because of reasons described in the earlier paragraph, analyzing machines with field solvers to identify fault signatures would be inordinately time consuming.

Describing electric machines as group coupled magnetic circuits provides another way of obtaining their operating characteristics. The circuit elements are usually resistances and inductances. Of the two, the latter are most difficult to compute because they vary with position of the rotating member as well as magnetic saturation.

The winding and modified winding function approach (WFA/MWFA) provides the necessary tool to compute these inductances. It provides a very computationally efficient way to estimate inductances from the machine winding and the air-gap data. Since the winding structure dictates the magnetomotive force inside a machine and the air-gap the bulk of the permeance;

flux, flux linkage, and hence flux linkage per ampere turn or inductance can be easily computed using this method. Even effects such as saturation, slots can be modeled by suitably modifying the air-gap permeance. Three-dimensional effects such as skewing and inclined rotor eccentricity can also be included. The WFA and MWFA are described next [1–3]. The analysis also makes the following assumptions:

1. Flux crosses the air-gap radially (axial flux is negligible).
2. Saturation is negligible.
3. Average core saturation is incorporated by using Carter's coefficient to adjust air-gap length.
4. Eddy current, friction, and windage losses are neglected.
5. The magnetic material has infinite permeance.
6. Slot effects are negligible.

3.2 Winding and Modified Winding Function Approaches (WFA and MWFA)

An elementary nonsalient motor with cylindrical stator and rotor is shown in Figure 3.1. The permeability of the stator and rotor iron cores is assumed to be infinite when compared to the permeability of the air-gap. The stator reference

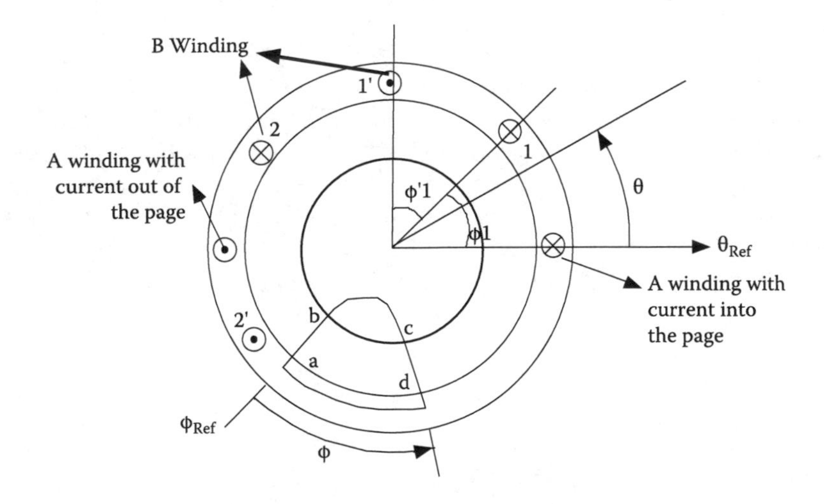

FIGURE 3.1
Elementary nonsalient uniform air-gap machine (From S. Nandi, "Fault Analysis for Condition Monitoring of Induction Motors," PhD disseration, Texas A&M University, May 2000.).

position, the angle φ, of the closed path *abcda* of Figure 3.1 is taken at an arbitrary point along the gap. Points *a* and *d* are located on the stator corresponding to angles 0 and φ, respectively, and points *b* and *c* are located on the rotor. In the case of smooth air-gap machines, the flux lines can be assumed to be radial and intersect with the rotor and stator at right angle. Using Gauss's law for magnetic fields, points *b* and *c* are uniquely defined since two flux lines can never originate from the same point if points *a* and *d* are fixed on the stator.

Consider the path *abcda* of Figure 3.1 for an arbitrary $0 < \varphi < 2\pi$, by Ampere's law

$$\oint_{abcda} H.dl = \int_S J.dS \tag{3.1}$$

where S is the surface enclosed by the path *abcda*. Since all the windings enclosed by the closed path carry the same current *i*, Equation (3.1) reduces to the following:

$$\oint_{abcda} H.dl = n(\phi,\theta).\, i \tag{3.2}$$

where H is the magnetic field intensity and dl is defined to be along the flux lines originating or terminating at two points of the closed path *abcda*. The function $n(\varphi,\theta)$ is called the *turns function* and represents the number of turns of the winding enclosed by the path *abcda*. In general, for a stationary coil it is only a function of φ. For a rotating coil it is assumed to be a function of φ and the rotor position angle θ. Turns carrying currents *i* into the page are considered positive while the turns carrying currents *i* out of the page are considered negative.

In terms of magnetomotive force (MMF) drops in a magnetic circuit, Equation (3.2) can be written as

$$F_{ab} + F_{bc} + F_{cd} + F_{da} = n(\phi,\theta).\, i \tag{3.3}$$

Since the iron is considered to be infinitely permeable, the MMF drops F_{bc} and F_{da} are negligible and Equation (3.3) reduces to

$$F_{ab}(0,\theta) + F_{cd}(\varphi,\theta) = n(\varphi,\theta).\, i \tag{3.4}$$

Gauss's law for magnetic field can be used to find an expression for the MMF drop at $\varphi = 0$, $F_{ab}(0,\theta)$, which is given by

$$\oint_S B.dS = 0 \tag{3.5}$$

where B represents the magnetic flux density and the surface integral is carried out over the boundary surface of an arbitrary volume. Taking the surface S to be a cylindrical volume located just inside the stator inner surface, Equation (3.5) can be written as

$$\int_0^{2\pi}\int_0^{l} \mu_0 H(\phi,\theta) r(dl)(d\phi) = 0 \tag{3.6}$$

where l is the axial stack length of the machine, r is the stator inner radius, μ_0 is the free space permeability, and θ is the angular position of the rotor with respect to stator. Since B does not vary with respect to the axial length, and MMF ($F(\varphi,\theta)$) is the product of radial length ($g(\varphi,\theta)$) and the magnetic field intensity ($H(\varphi,\theta)$), then

$$\int_0^{2\pi} \frac{F(\phi,\theta)}{g(\phi,\theta)} d\phi = 0 \tag{3.7}$$

Dividing Equation (3.4) by the *air-gap function* $g(\varphi,\theta)$, and then integrating from 0 to 2π yields

$$\int_0^{2\pi} \frac{F_{ab}(0,\theta) + F_{cd}(\phi,\theta)}{g(\phi,\theta)} d\phi = \int_0^{2\pi} \frac{n(\phi,\theta)}{g(\phi,\theta)} i \, d\phi \tag{3.8}$$

Since the second term of the left-hand side is zero as found from Gauss's law, Equation (3.8) will reduce to the following:

$$F_{ab}(0,\theta) = \frac{1}{2\pi < g^{-1}(\phi,\theta) >} \int_0^{2\pi} n(\phi,\theta)\, g^{-1}(\phi,\theta)\, i \, d\phi \tag{3.9}$$

where $< g^{-1}(\varphi,\theta) >$ is the average value of the inverse gap function. Substituting Equation (3.9) in Equation (3.4) and solving for $F_{cd}(\varphi,\theta)$ yields

$$F_{cd}(\phi,\theta) = \left[n(\phi,\theta) - \frac{1}{2\pi < g^{-1}(\phi,\theta) >} \int_0^{2\pi} n(\phi,\theta)\, g^{-1}(\phi,\theta)\, d\phi \right] . i \tag{3.10}$$

From the last equation, the *winding* and *modified winding function*, in general, can be defined respectively as follows:

$$N(\phi,\theta) = n(\phi,\theta) - < n(\phi,\theta) > \tag{3.11}$$

$$M(\phi,\theta) = n(\phi,\theta) - < M(\phi,\theta) > \tag{3.12}$$

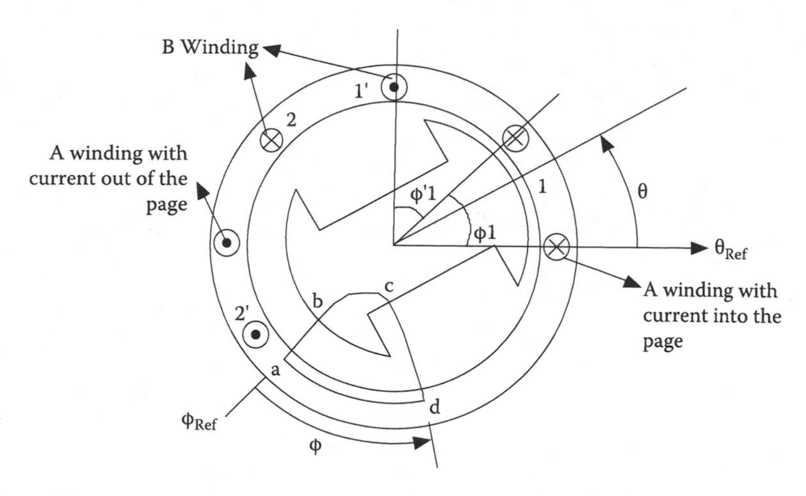

FIGURE 3.2
Elementary salient pole machine with eccentric rotor. (From N.A. Al-Nuaim and H.A. Toliyat, "A novel method for modeling dynamic air-gap eccentricity in synchronous machines based on modified winding function theory," *IEEE Transactions on Energy Conversion*, vol. 13, no. 2, pp. 156–162, June 1998. With permission).

where

$$< n(\phi,\theta) > = \frac{1}{2\pi}\int_0^{2\pi} n(\phi,\theta)\ d\phi \tag{3.13}$$

$$< M(\phi,\theta) > = \frac{1}{2\pi < g^{-1}(\phi,\theta) >}\int_0^{2\pi} n(\phi,\theta)\, g^{-1}(\phi,\theta)\ d\phi \tag{3.14}$$

When the air-gap is symmetric, Equation (3.11) is applicable. In case of machines with eccentric air-gap or saliency, however, Equation (3.12) is applicable in general as the air-gap becomes a function of φ and θ. An example of such an eccentric machine is shown in Figure 3.2. Also note that while average value of the turns functions as given by Equation (3.13) and Equation (3.14) are not explicitly dependent on φ, the reference point chosen has a definite influence on it.

Example 3.1

a. Compute the winding function for the full pitch N_s turn stator winding ($N_s = 100$) shown in Figure 3.3. The origin has been chosen so as to make the fundamental component of winding function a cosine function. Assume uniform air-gap of $g(\phi,\theta) = g_0$.

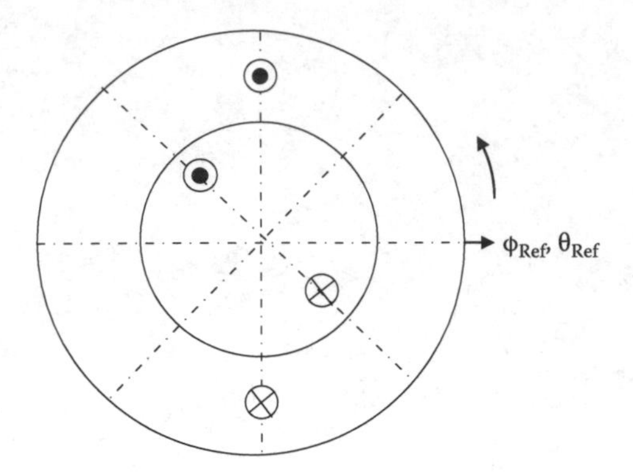

FIGURE 3.3
Elementary N_s turn coil on stator, and N_r turn coil on rotor.

b. Compute the modified winding function for the same winding when the air-gap is eccentric and is given by $g(\phi,\theta) = g_0 - \delta g_0 \cos\phi$ with $\delta = 0.5$.

Solution

a. The turns function for the winding given in Figure 3.3 is first computed using Equation (3.2). The winding function for this winding can be computed using Equation (3.11) as $N_s(\phi,\theta) = \frac{2N_s}{h\pi}\sum_{h=1,3,5...}^{\infty}(-1)^{\frac{h-1}{2}}\cos h\phi$. The turns function and winding function for different values of φ are shown in Figure 3.4 (top) and Figure 3.4 (middle) respectively. Using Equation (3.13) it is easy to show that $< n(\phi,\theta) > \; = \; -\frac{N_s}{2}$. However if the dots and crosses in

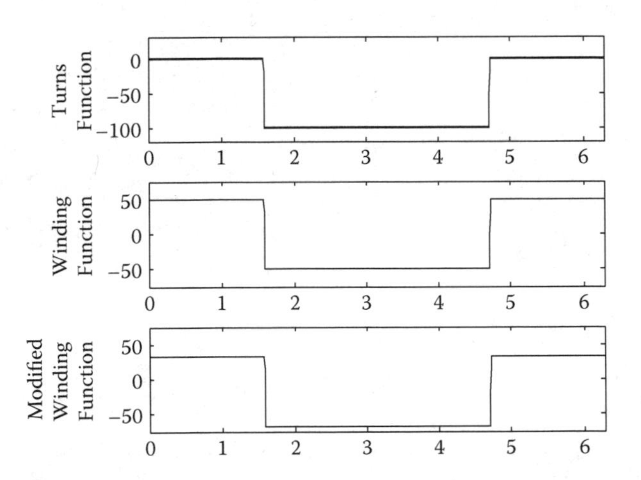

FIGURE 3.4
(From top) Turns function, winding function, and modified winding function.

the conductors are swapped or the φ_{ref} and θ_{ref} points are shifted anticlockwise by 90°, $< n(\phi,\theta) > = \frac{N_s}{2}$. This explains the implicit position dependence of the $< n(\phi,\theta) >$

b. With $g^{-1}(\phi,\theta) = \frac{1}{g_0 - \delta g_0 \cos\phi} \approx A_1 + A_2\cos(\phi)$, where $A_1 = \frac{1}{g_0\sqrt{1-\delta^2}}$; $A_2 = \frac{2}{g_0\sqrt{1-\delta^2}}\left(\frac{1-\sqrt{1-\delta^2}}{\delta}\right)$. Using Equations (3.11) to (3.14) it can be shown that $< M(\theta) > = \frac{N_s A_2}{\pi A_1} - \frac{N_s}{2}$. Then using Equation (3.11) and Equation (3.12) $M(\phi,\theta) = N(\phi,\theta) - \frac{N_s A_2}{\pi A_1}$. Figure 3.4 (bottom) shows the modified winding function.

3.3 Inductance Calculations Using WFA and MWFA

In the previous section, the relative permeability of the iron was assumed to be infinity; that is, the MMF drop in the iron was neglected. Hence, the MMF distribution of machine windings in the air-gap can simply be found by the product of WFA or MWFA calculated from Equation (3.11) or Equation (3.12) and the current flowing in the winding.

As shown in Figure 3.1, windings A and B are located in the air-gap and could be associated with either the rotor or the stator. The mutual inductance of winding B due to current i_A flowing in winding A is to be calculated. Winding B is arranged arbitrarily in the air-gap and for demonstration is assumed to have two coil sides, 1–1′ and 2–2′, with different turns distribution in the air gap. The reference angle φ cannot be selected freely and should be the same reference position that has been used previously to calculate the modified winding function $N_A(\varphi,\theta)$ or $M_A(\varphi,\theta)$.

The following derivations have been shown with MWFA. Similar results can be obtained with WFA using $N_A(\varphi,\theta)$ instead of $M_A(\varphi,\theta)$ and noting that the air-gap and the inverse air-gap functions can be defined as constants.

The MMF distribution in the air gap due to current i_A can be calculated as follows:

$$F_A(\phi,\theta) = M_A(\phi,\theta).i_A \tag{3.15}$$

It is known that the flux in a magnetic circuit is the product of the MMF (F) and the permeance (P) of the flux path. Thus,

$$\Phi = F.P \tag{3.16}$$

and the permeance is given by

$$P = \frac{\mu A}{l} \tag{3.17}$$

where μ is the permeability, A is the cross-sectional area, and l is the length of the magnetic path. The differential flux across the gap through a differential volume of length $g(\varphi,\theta)$ and cross-sectional area of $r\ l.d\varphi$ from the rotor to the stator is

$$d\Phi = F_A(\phi,\theta)\mu_o\, r l\, g^{-1}(\phi,\theta)d\phi \tag{3.18}$$

The flux linking the coil sides 1–1′ of winding B can be calculated using the following integration

$$\Phi_{1-1'} = \mu_0\, r l \int_0^{2\pi} n_{B1}(\phi,\theta)\ F_A(\phi,\theta)\ g^{-1}(\phi,\theta)\ d\phi \tag{3.19}$$

where $n_{B1}(\varphi,\theta)$ is equal to the number of turns of coil sides 1–1′ between the reference angles φ_1 and φ'_1 of Figure 3.1 and zero otherwise. Coil side 1′ is the return path for coil side 1. Continuing the process of calculating the flux linking the other coil sides of winding B and in general for any set of coil sides k–k' the flux linkage is

$$\Phi_{k-k'} = \mu_0\, r l \int_0^{2\pi} n_{Bk}(\phi,\theta)\ F_A(\phi,\theta)\ g^{-1}(\phi,\theta)\ d\phi \tag{3.20}$$

where $n_{Bk}(\varphi,\theta)$, $F_A(\varphi\ ,\theta)$, and $g^{-1}(\varphi,\theta)$ must have the same position reference φ. The total flux linking winding B due to current in winding A can be defined as follows

$$\lambda_{BA} = \sum_{k=1}^{q_i} \Phi_{k-k'} = \mu_0\, r l \left[\sum_{k=1}^{q_i} \int_0^{2\pi} n_{Bk}(\phi,\theta) F_A(\phi,\theta) g^{-1}(\phi,\theta) d\phi \right] \tag{3.21}$$

or

$$\lambda_{BA} = \mu_0\, r l \int_0^{2\pi} \left[\sum_{k=1}^{q_i} n_{Bk}(\phi,\theta) \right] F_A(\phi,\theta)\, g^{-1}(\phi,\theta)\, d\phi \tag{3.22}$$

where

$$n_B(\phi,\theta) = \left[\sum_{k=1}^{q_i} n_{Bk}(\phi,\theta) \right] \tag{3.23}$$

is the turns function for the winding B assuming that the q_i coil sides are connected in series.

The mutual inductance L_{BA} of winding B due to the current i_A in winding A would be

$$L_{BA} = \frac{\lambda_{BA}}{i_A} = \mu_0 \, r l \int_0^{2\pi} n_B(\phi,\theta)\; M_A(\phi,\theta)\; g^{-1}(\phi,\theta)\; d\phi \tag{3.24}$$

Using the same process, the magnetizing inductance of winding A can be defined as

$$L_{AA} = \mu_0 \, r l \int_0^{2\pi} n_A(\phi,\theta)\; M_A(\phi,\theta)\; g^{-1}(\phi,\theta)\; d\phi \tag{3.25}$$

Integrations such as Equation (3.24) and Equation (3.25) can be performed by expressing the turns, modified winding, and inverse air-gap functions as Fourier series. This way, the effects of space and air-gap permeance harmonics can be included in the simulation studies. Stator and rotor slot effects can also be included by adding sinusoidal functions with frequency components as a function of the number of slots. Even saturation effects and rotor skewing can be included. Figure 3.5 shows an inductance profile computed using saturation effects. The inverse air-gap in this case is defined as

$$g_s^{-1}(\phi,\theta_f\,) \approx \frac{1}{g'}[1 + k_{gsat} Cos\{2(p\phi - \theta_f)\}] \tag{3.26}$$

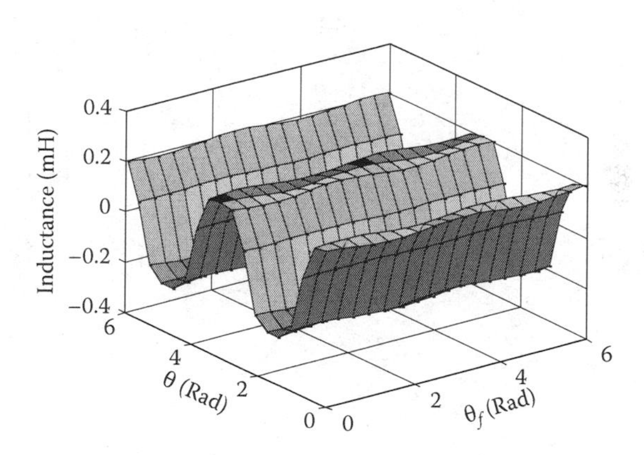

FIGURE 3.5
Simulated inductance profile between stator phase *a* and rotor loop *1* of an induction motor. It is assumed that saturation causes a 6% increase in air gap at tip of the air-gap flux vector (k_{gsat} = 0.06). θ is the rotor position and θ_f is the air-gap flux vector position. The rotor is skewed by about 42% of one rotor slot. (From S. Nandi, "A detailed model of induction machines with saturation extendable for fault analysis," *IEEE Transactions on Industry Applications*, vol. 40, no. 5, pp. 1302–1309, September/October 2004. With permission.).

where θ_f is the position of the air-gap flux measured from phase a axis, $p$is the number of fundamental pole pairs, g' is the modified average value of the air-gap in the presence of saturation and k_{gsat} is the saturation factor derived on the assumption that saturation modulates the air-gap permeance at twice the frequency of the air-gap flux density wave.

A question often arises as to whether in the presence of air-gap nonuniformity the reciprocity theorem holds for mutual inductance; that is, if

$$L_{AB} = L_{BA} \tag{3.27}$$

The following derivation [5] shows that this is indeed true. Substituting $M_A(\phi,\theta)$ in Equation (3.24) by its equivalent of Equation (3.12) leads to

$$L_{BA} = \mu_0 rl \int_0^{2\pi} n_B(\phi,\theta)(n_A(\phi,\theta) - \langle M_A(\phi,\theta)\rangle) g^{-1}(\phi,\theta)\ d\phi \tag{3.28}$$

or

$$L_{BA} = \mu_0 rl \begin{bmatrix} \int_0^{2\pi} n_B(\phi,\theta) n_A(\phi,\theta)\ g^{-1}(\phi,\theta)\ d\phi \\ -\int_0^{2\pi} n_B(\phi,\theta)\ \langle M_A(\phi,\theta)\rangle g^{-1}(\phi,\theta)\ d\phi \end{bmatrix} \tag{3.29}$$

Since $< M_A(\phi,\theta) >$ is a constant, it can be taken outside the integral sign. Thus

$$L_{BA} = \mu_0 rl \begin{bmatrix} \int_0^{2\pi} n_B(\phi,\theta) n_A(\phi,\theta)\ g^{-1}(\phi,\theta)\ d\phi \\ -\langle M_A(\phi,\theta)\rangle \int_0^{2\pi} n_B(\phi,\theta)\ g^{-1}(\phi,\theta)\ d\phi \end{bmatrix} \tag{3.30}$$

or

$$L_{BA} = \mu_0 rl \begin{bmatrix} \int_0^{2\pi} n_A(\phi,\theta) n_B(\phi,\theta)\ g^{-1}(\phi,\theta)\ d\phi \\ -2\pi \langle M_A(\phi,\theta)\rangle \langle M_B(\phi,\theta)\rangle \langle g^{-1}(\phi,\theta)\rangle \end{bmatrix} \tag{3.31}$$

Equation (3.31) can be obtained by substituting $\int_0^{2\pi} n_B(\phi,\theta)g^{-1}(\phi,\theta)\ d\phi$ with $2\pi\langle M_B(\phi,\theta)\rangle\langle g^{-1}(\phi,\theta)\rangle$ in Equation (3.30), a relation derivable from inspection of Equation (3.14).

Similar procedure will show that inductance of coil A due to current in coil B, given by,

$$L_{AB} = \mu_0\, r l \int_0^{2\pi} n_A(\phi,\theta)\ M_B(\phi,\theta)\ g^{-1}(\phi,\theta)\ d\phi \tag{3.32}$$

is also equal to Equation (3.31). This completes the proof of Equation (3.27).

Recently, axial air-gap asymmetry has also been accounted for calculating inductances using MWFA approach. The inductance is then computed as

$$L_{AB} = \mu_o r \left[\begin{array}{c} \displaystyle\int_0^l \int_0^{2\pi} n_A(\phi,\theta,y) n_B(\phi,\theta,y) g^{-1}(\phi,\theta,y)\, d\phi\, dy \\ \displaystyle -2\pi \int_0^l \langle M_A(\phi,\theta,y)\rangle\langle M_B(\phi,\theta,y)\rangle\langle g^{-1}(\theta,y)\rangle\, dy \end{array} \right] \tag{3.33}$$

Also, $L_{AB} = L_{BA}$ as before.

The modified winding function in this case has been defined as

$$M(\phi,\theta,y) = n(\phi,\theta,y) - \langle M(\phi,\theta,y)\rangle \tag{3.34}$$

Here $n(\varphi,\theta,y)$ is the turns function of the winding, and the $\langle M(\phi,\theta,y)\rangle$ is the average value of the modified winding function, which can be expressed as

$$\langle M(\phi,\theta,y)\rangle = \frac{1}{2\pi\langle g^{-1}(\phi,\theta,y)\rangle} \int_0^{2\pi} n(\phi,\theta,y) g^{-1}(\phi,\theta,y)\, d\phi \tag{3.35}$$

Where

$$\langle g^{-1}(\phi,\theta,y)\rangle = \frac{1}{2\pi} \int_0^{2\pi} g^{-1}(\phi,\theta,y)\, d\phi \tag{3.36}$$

is the average part of $g^{-1}(\phi,\theta,y)$.

Example 3.2

a. Compute the mutual inductance between the full pitch N_s turn stator winding and the full pitch N_r turn rotor winding (N_r = 50) shown in Figure 3.3. Assume uniform air-gap of $g(\phi,\theta) = g_0 = 0.5$ mm. Assume L = 100 mm and r = 25 mm.
b. Compute the same when the air-gap is static eccentric and is given by $g(\phi,\theta) = g_0 - \delta g_0 \cos\phi$ with δ=0.5. L and r are same as before.

Solution

a. With uniform air-gap, Equation (3.24) reduces to $L_{BA} = \frac{\lambda_{BA}}{i_A} = \frac{\mu_0 r l}{g_0} \int_0^{2\pi} N_B(\phi,\theta) N_A(\phi,\theta) \quad d\phi$. Since the winding function of winding A is periodic with zero average value.

With

$$N_B(\phi,\theta) = N_s(\phi,\theta) = \frac{2N_s}{h\pi} \sum_{h=1,3,5...}^{\infty} (-1)^{\frac{h-1}{2}} \cos h\phi$$

and

$$N_A(\phi,\theta) = N_r(\phi,\theta) = \frac{2N_r}{h\pi} \sum_{h=1,3,5...}^{\infty} (-1)^{\frac{h-1}{2}} \cos h(\phi - \theta),$$

one can compute L_{sr} as

$$L_{sr} = \frac{\mu_0 rl}{g} \frac{4N_s N_r}{\pi} \sum_{h=1,3,5..}^{\infty} \frac{\cos h\theta}{h^2}.$$

The plot of L_{BA} versus θ is shown in Figure 3.6 (top).

b. Using Equation (3.24) and A_1, A_2 as shown in *Example 3.1* it can be shown that in presence of static eccentricity

$$L_{sre} = \mu_0 rl \frac{2N_s N_r}{\pi} \left(-\frac{A_2^2}{A_1} \cos\theta + 2A_1 \sum_{h=1,3,5..}^{\infty} \frac{\cos h\theta}{h^2} \right).$$

The plot of L_{sre} versus θ is as shown in Figure 3.6 (bottom).

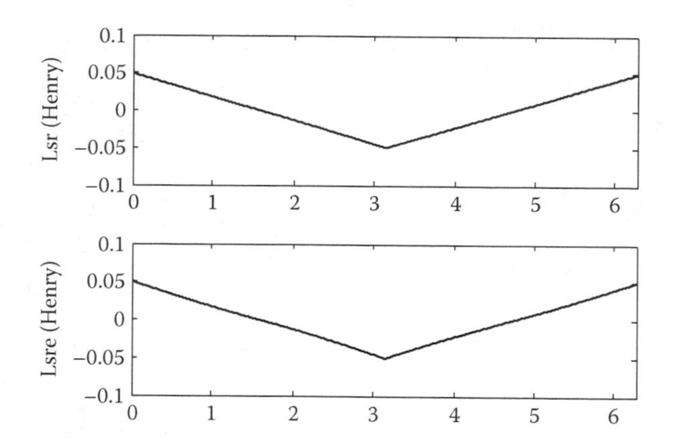

FIGURE 3.6
Stator–rotor mutual inductance with uniform (top) and static eccentric air-gap (bottom).

3.4 Validation of Inductance Calculations Using WFA and MWFA

The inductance generated this way is the key toward modeling electric machines. However, an obvious question that might arise in the reader's mind is regarding the veracity and the accuracy of WFA and MWFA based techniques, particularly in view of the overly simplified assumptions made in Section 3.2.

Although it is almost impossible to measure some of these inductances in a real machine, realistic and close results can be obtained by solving Maxwell equations provided the motor structural details are accurately known. This can easily be done using the various finite-element (FE)-based Maxwell equation solvers available. "Maxwell," from Ansoft Corporation is one of the more popular FE-based field distribution solvers currently available commercially. The FE-based validation examples of the inductance profiles generated using WFA and MWFA have been created using this package.

Figure 3.7 shows the comparative plots for a healthy induction machine and mixed eccentric induction machine using WFA/MWFA and FE. The values used were 15% static eccentricity and 5% dynamic eccentricity. The ripples in the FE plots are due to rotor and stator slotting that was not modeled in WFA/MWFA.

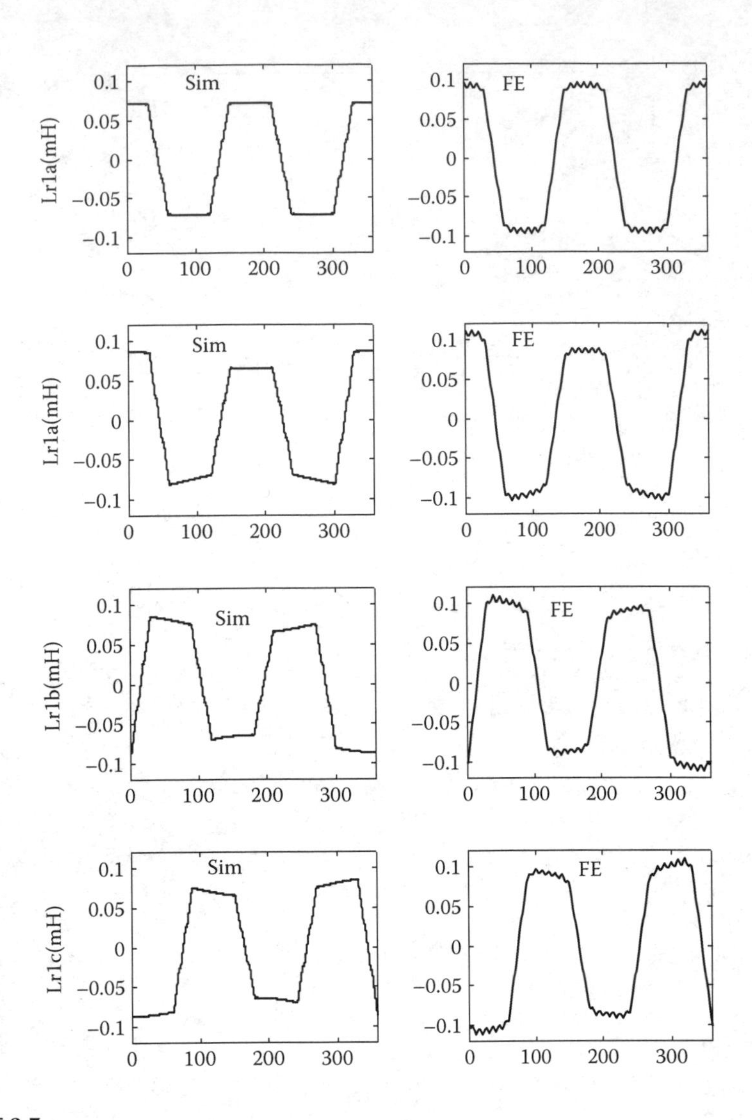

FIGURE 3.7
Simulated mutual inductance between rotor loop *1* and stator phases of an induction obtained from WFA/MWFA and FE analysis. From top: Healthy machine (*a* phase); mixed eccentric machine (phases *a*, *b*, and *c*,) with 15% SE and 5% DE. (From S. Nandi, R.M. Bharadwaj, and H.A. Toliyat, "Mixed eccentricity in three phase induction machines: Analysis, simulation and experiments," *Proceedings of the 37th IAS Annual Meeting*, vol. 3, pp. 1525–1532, October 2002. With permission.)

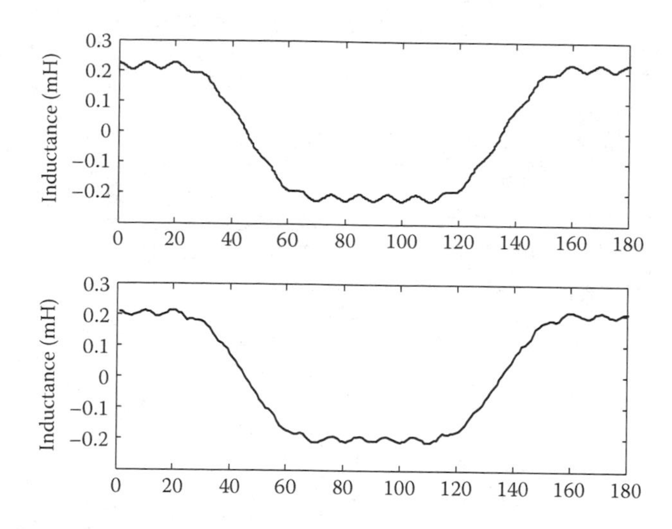

FIGURE 3.8
Simulated mutual inductance profile between stator phase *a* and rotor loop *1* of an induction motor including slot effects using (MWFA*)* (top) and FE (bottom). (From S. Nandi, "Modeling of induction machines including stator and rotor slot effects," *IEEE Transactions on Industry Applications,* vol. 40, no. 4, pp. 1058–1065, July/August 2004. With permission.)

Figure 3.8 shows the comparative plots for machines whose slot effects have been modeled using MWFA. The induction machine modeled had 36 stator slots and 28 rotor slots. The slot effects using MWFA have been defined by the following inverse air-gap function:

$$g_{sr}^{-1}(\phi,\theta) = \frac{1}{g}\left[\begin{aligned} &\frac{1}{k_{c1}k_{c2}} - \frac{a_1}{k_{c2}}Cos(S\phi) - \frac{b_1}{k_{c1}}Cos\{R(\phi-\theta)\} \\ &+\frac{a_1 b_1}{2}Cos\{(S-R)\phi + R\theta\} + \frac{a_1 b_1}{2}Cos\{(S+R)\phi - R\theta\}. \end{aligned}\right] \tag{3.37}$$

where k_{c1}, k_{c2} stator and rotor Carter's coefficients respectively; a_1, b_1 are parameters dependent on slot and other machine geometry (1/m); S and R are the number of stator and rotor slots, respectively.

Figure 3.9 shows results where MWFA has been extended to include 3D effect such as axial inclination. Figure 3.10 illustrates comparative inductance profile for a salient pole synchronous machine with dynamic eccentricity.

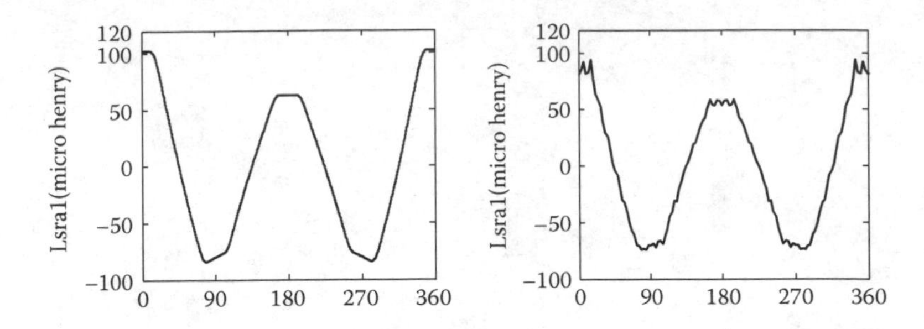

FIGURE 3.9
Simulated mutual inductance profile between stator phase *a* and rotor loop *1* of an induction motor using MWFA (left) and 3D FE (right) with 0% static eccentricity on one end of the shaft and 50% static eccentricity at the other end. (From X. Li, Q. Wu, and S. Nandi, "Performance analysis of a 3-phase induction machine with inclined static eccentricity," *IEEE Transactions on Industry Applications*, vol. 43, no. 2, pp. 531–541, March/April 2007. With permission.)

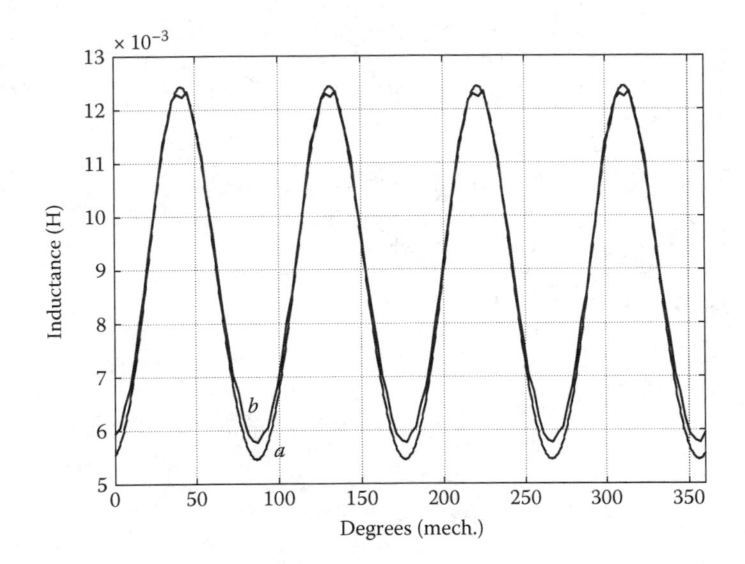

FIGURE 3.10
Stator *a* phase magnetizing inductance of a synchronous motor using MWFA (a) and FE method (b) with 25% dynamic air-gap eccentricity. (From H.A. Toliyat and N.A. Al-Nuaim, "Simulation and detection of dynamic air-gap eccentricity in salient-pole synchronous machines," *IEEE Transactions on Industry Applications*, vol. 35, no. 1, pp. 86–93, January/February 1999.)

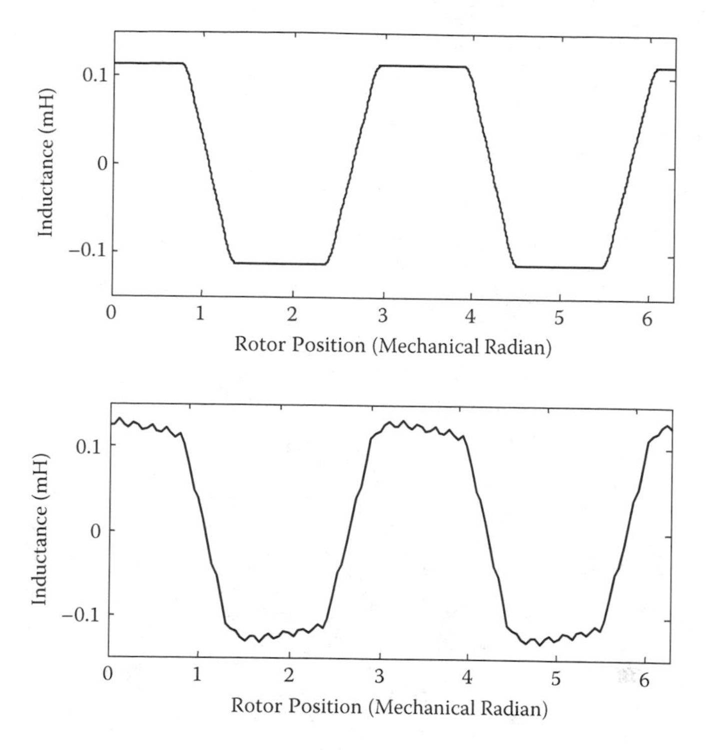

FIGURE 3.11
Mutual inductance of stator *a* phase to one rotor loop of a healthy RSM using MWFA (top) and FE (bottom). (From P. Neti, "Stator Fault Analysis of Synchronous Machines," PhD dissertation, University of Victoria, December 2007. With permission.)

Figure 3.11 presents the inductance profile of a synchronous reluctance machine using MWFA with comparative FE generated inductance profile. Figure 3.12 shows a similar comparison of inductance profiles of a standard synchronous machine (SM).

These results clearly demonstrate the power of WFA/MWFA where inductance profiles could be computed at much higher resolution but at a fraction of time required for FE methods and without sacrificing much accuracy. It is shown by Ilamparithi and Nandi that a commercial FE simulation package could take around 50 hours to simulate 1.5 seconds of steady-state run of a 3 hp, 44 rotor bar induction motor compared to 4 minutes of running time for a MWFA-based coupled-inductive circuit MATLAB code while generating similar spectral characteristics [10]. Table 3.1 provides the detailed results.

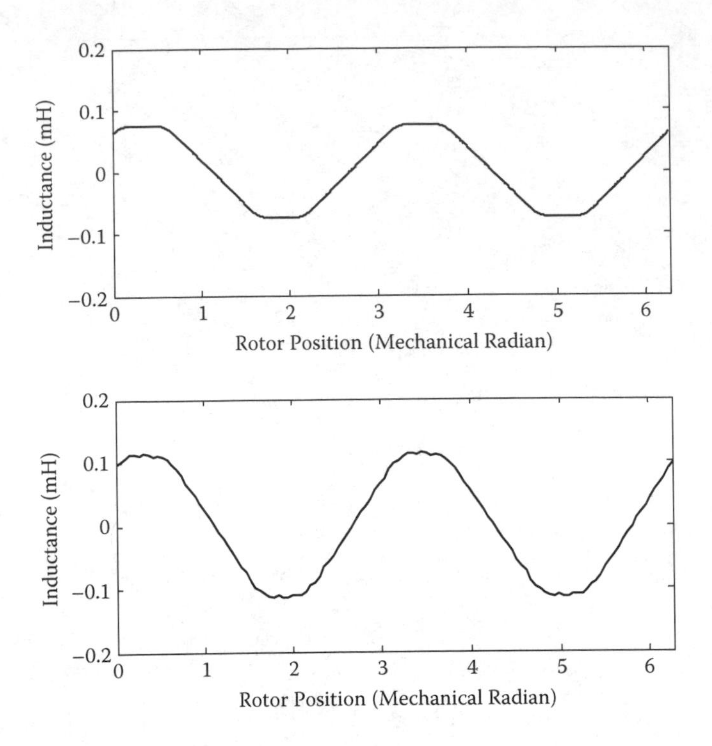

FIGURE 3.12
Mutual inductance of stator *a* phase to one rotor loop of a healthy SM using MWFA (top) and FE (bottom). (From P. Neti, "Stator Fault Analysis of Synchronous Machines," PhD dissertation, University of Victoria, December 2007. With permission.)

TABLE 3.1
Normalized (with Respect to Fundamental 60 Hz Which Is At 0 db) Magnitude of Fault Specific Frequency Components for Different Eccentric Conditions

Fault Indicating Component	MWFA (dB)	FE (dB)
50% SE (1388 Hz)	–39.66	–40.81
50% DE (1358 Hz)	–38.88	–40.87
ME (25% SE + 25% DE)		
(29.5Hz)	–45.5	–48.85
(89.5 Hz)	–48.28	–50.89
(1358 Hz)	–45.74	–46.84
(1388 Hz)	–47.02	–47.08

References

[1] T.A. Lipo, *Analysis of Synchronous Machine*, Wisconsin Power Electronics Research Center, University of Wisconsin, 2008.
[2] S. Nandi, "Fault Analysis for Condition Monitoring of Induction Motors," PhD dissertation, Texas A&M University, May 2000.
[3] N.A. Al-Nuaim and H.A. Toliyat, "A novel method for modeling dynamic air-gap eccentricity in synchronous machines based on modified winding function theory," *IEEE Transactions on Energy Conversion*, vol. 13, no. 2, pp. 156–162, June 1998.
[4] S. Nandi, "A detailed model of induction machines with saturation extendable for fault analysis," *IEEE Transactions on Industry Applications*, vol. 40, no. 5, pp. 1302–1309, September/October 2004.
[5] S. Nandi, R.M. Bharadwaj, and H.A. Toliyat, "Mixed eccentricity in three phase induction machines: Analysis, simulation and experiments," *Proceedings of the 37th IAS Annual Meeting*, vol. 3, pp. 1525–1532, October 2002.
[6] S. Nandi, "Modeling of induction machines including stator and rotor slot effects," *IEEE Transactions on Industry Applications*, vol. 40, no. 4, pp. 1058–1065, July/August 2004.
[7] X. Li, Q. Wu, and S. Nandi, "Performance analysis of a 3-phase induction machine with inclined static eccentricity," *IEEE Transactions on Industry Applications*, vol. 43, no. 2, pp. 531–541, March/April 2007.
[8] H.A. Toliyat and N.A. Al-Nuaim, "Simulation and detection of dynamic air-gap eccentricity in salient-pole synchronous machines," *IEEE Transactions on Industry Applications*, vol. 35, no. 1, pp. 86–93, January/February 1999.
[9] P. Neti, "Stator Fault Analysis of Synchronous Machines," PhD dissertation, University of Victoria, December 2007.
[10] T. Ilamparithi and S. Nandi, "Comparison of results for eccentric cage induction motor using finite element method and modified winding function approach," *PEDES 2010* Conference, pp. 1–7, New Delhi, India, December 20–23, 2010,

4

Modeling of Electric Machines Using Magnetic Equivalent Circuit Method

Homayoun Meshgin-Kelk, Ph.D.
Tafresh University

4.1 Introduction

The magnetic equivalent circuit (MEC) method introduces another approach for modeling electric machines. In fact, the approach can be considered as a reduced order finite element (FE) method. By taking into account approximately accurate machine geometry, stator and rotor slots effects, skewing, type of winding connections, stator and rotor leakages, and linear or nonlinear magnetic characteristics of machine coress it is a more accurate method with respect to the winding function approach (WFA). Therefore it can be helpful for design engineers and also it may be applied to find and to study more reliable algorithms for fault-detection strategies. Neglecting core property MEC is very similar to WFA.

Although WFA is based on calculating machine inductances, the magnetic equivalent circuit method can be used in two ways, indirect and one direct one. In the indirect way where linear magnetic core is considered, it may be applied to calculate machine inductances as the first step in analyzing the machine performances. Since magnetic properties of core parts can be incorporated in calculating machine inductances, it can provide a more accurate way to calculate these inductances. On the other hand, it may be applied directly without calculating machine inductances to analyze machine behavior in most conditions. MEC usually provides a deep understanding about effects of the machine geometry and design data on its parameters and performance.

Most conventional machines, such as induction and synchronous machines, are divided into three main parts: the stator, the rotor, and the air-gap. The stator consists of a stator core and stator windings, and the rotor consists of a rotor core and rotor windings. Stator or rotor cores are divided into yoke and teeth. Stator windings are located in the stator slots and rotor windings are located in the rotor slots. Figure 4.1 shows a portion of the main parts of a typical electric machine. Several magnetic flux lines are also shown. The

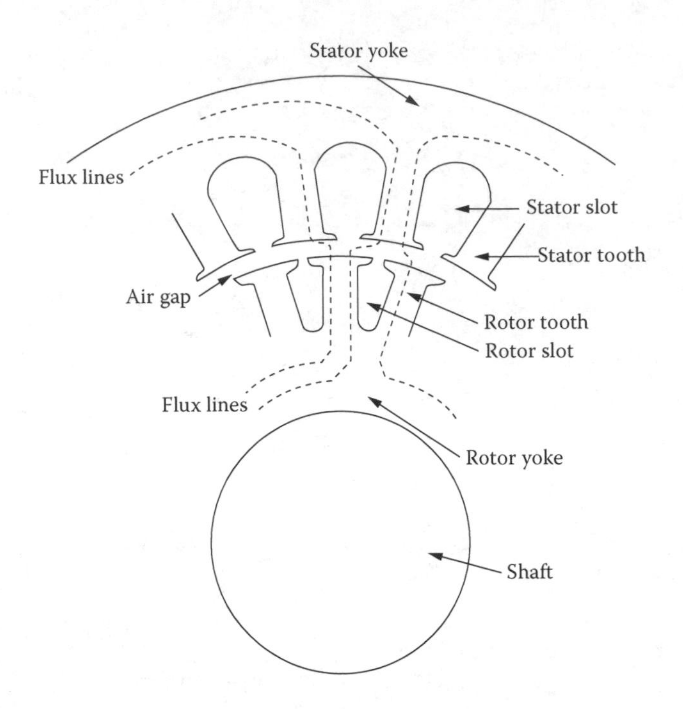

FIGURE 4.1
A portion of the main parts of a typical electric machine.

flux lines that link both the stator and the rotor windings are the useful flux linkages, and the flux lines that link only the stator or the rotor windings are called leakage fluxes. In Figure 4.1, three lines of fluxes are shown.

An electric machine is assumed to be a quasi-stationary device; that is, any change of current that builds the flux is followed by an immediate change of flux. In other words, the time needed for an electromagnetic wave to pass through the machine is negligible compared to the period of the wave. Such a space may be partitioned into flux tubes. The flux tubes are the basis of the magnetic equivalent circuit method. A flux tube is a geometrical space in which all lines of flux are perpendicular to their bases and no lines of flux cut their sides. Lines of equal magnetic scalar potential, u, are perpendicular to lines of flux,φ. Therefore the bases of a flux tube are equipotential planes. Magnetic scalar potential, u, is a very useful quantity in magnetic equivalent circuit theory. But it has no physical meaning like the electrical scalar potential, v.

Lines of constant scalar magnetic potential lie perpendicular to the H vector. Scalar magnetic potential [6,7] is defined by

$$\vec{H} = -\nabla u \tag{4.1}$$

The electrical scalar potential, v, in an electrostatic field is defined by

$$\vec{E} = -\nabla v \tag{4.2}$$

A similarity exists between the definitions of these potentials. If the reluctance of a flux tube with magnetic scalar potentials u_1 and u_2 at their bases is R, and the flux through it is φ, then

$$F = u_2 - u_1 = \phi.R \tag{4.3}$$

which states that $u_2 - u_1$ is the magnetomotive force (MMF) drop on the reluctance R.

To construct the magnetic equivalent circuit of an electric machine, all yoke parts, teeth, slots, air-gap tubes between the stator teeth and rotor teeth, and windings are modeled. If all teeth and yokes reluctances and all slot leakages are neglected and only air-gap permeances are considered, then the magnetic equivalent circuit approach is the same as the classic WFA. In this case, there exist MMF drops across the air-gap permeances due to currents flowing in the windings.

Currents flowing in the windings are the MMF sources in the magnetic equivalent circuit of an electric machine. These MMF sources are placed in the teeth. A reluctance/permeance and an unknown flux are assigned to each yoke part or each tooth. Slots are modeled by their leakage permeances. In the air-gap, a flux tube is defined when any of the stator teeth come face to face with any rotor teeth. A permeance is assigned to each air-gap flux tube. Geometries of these flux tubes vary due to the rotation of rotor teeth with respect to stator teeth and also depend on the air-gap length between any two facing stator and rotor teeth. Any air-gap asymmetry will have an effect on the height of these flux tubes.

To find the magnetic equivalent circuit of an electric machine we need first to compute the permeance between any two pairs of face-to-face stator and rotor teeth. Figure 4.2 shows three positions of *j*th rotor tooth with respect to the fixed *i*th stator tooth. The permeance between these two teeth is called G_{ij}. At positions *a* and *c*, G_{ij} is zero because there is no flux tube (no crossing flux) between these two teeth. If the widths of these teeth are equal, the G_{ij} reaches its maximum only at position *b*.

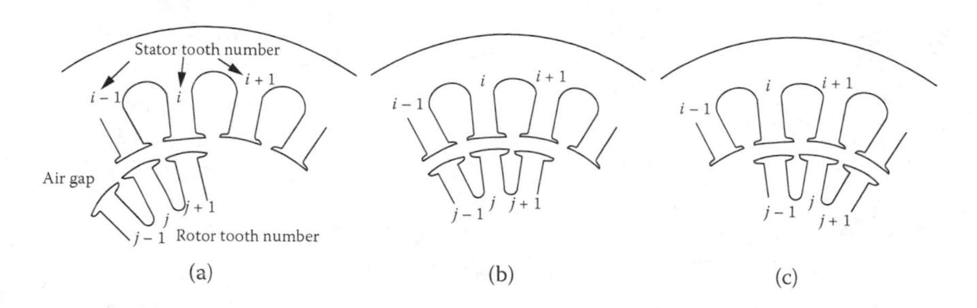

FIGURE 4.2
Three positions of *j*th rotor tooth with respect to the fixed *i*th stator tooth.

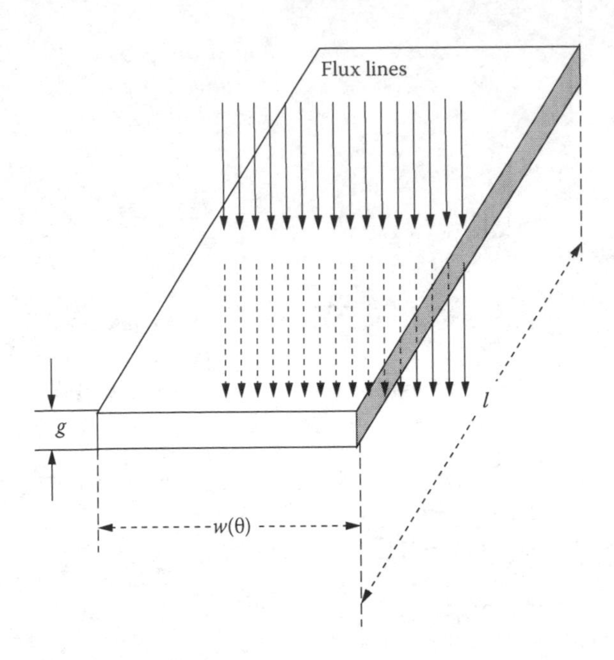

FIGURE 4.3
A flux tube with rectangular bases between *j*th rotor tooth and *i*th stator tooth for nonskewed stator and rotor slots.

For un-skewed stator and rotor slots, the bases of air-gap flux tubes are rectangles. In this case, the flux tubes have a general shape as shown in Figure 4.3 and the air-gap permeance G_{ij} can be calculated by

$$G_{ij} = \mu_0 \frac{l.w(\theta)}{g} \tag{4.4}$$

where l is the length, $w(\theta)$ is the width, and g is the height of the flux tube. $w(\theta)$ varies with the angular position of the stator tooth i and rotor tooth j. The air-gap length between the stator tooth i and the rotor tooth j, g, is constant for a symmetric air-gap. For a nonsymmetric air-gap, g is a function of θ.

If the widths of these teeth are not equal, the maximum of G_{ij} occurs during an interval. Figure 4.4 shows the air-gap permeance for two conditions. It is necessary to note that due to the fringing flux, the real shape of G_{ij} is smoother than what are shown in Figure 4.4.

If either the stator slots or the rotor slots, or both, are skewed, then the bases of the air-gap flux tubes will not be rectangles any more. Depending on the width of the stator and the rotor teeth and the amount of skewing and the angular position of rotor teeth to stator teeth, the bases of flux tubes have different shapes. Figure 4.5. shows the different shapes of bases for different

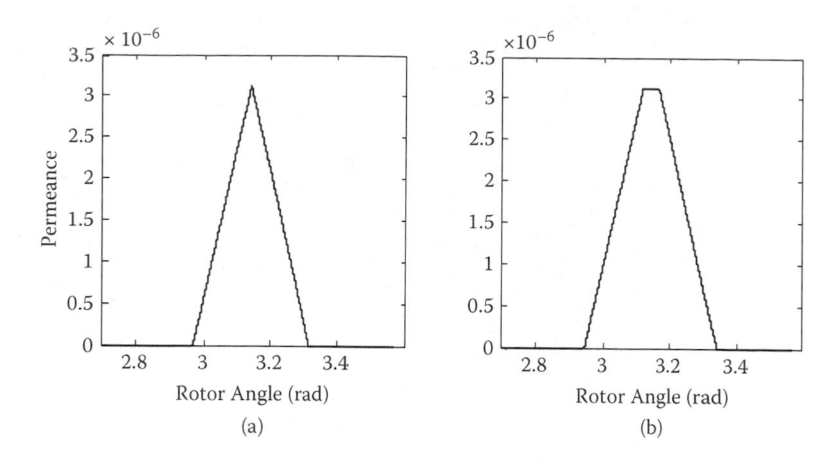

FIGURE 4.4
Air-gap permeance for two conditions: (a) equal width of stator and rotor teeth, (b) different width of stator and rotor teeth.

positions of a typical skewed rotor tooth with respect to a typical rectangular stator tooth. Shaded areas are the bases of flux tube as the rotor tooth moves.

Skewing is a method to get better performance of an electric machine. By skewing either the stator or the rotor, the shape of air-gap permeance function and its space derivative are smoother. The derivative of air-gap permeance G_{ij} when multiplied by the square of the MMF drop over the element G_{ij} gives the value of electromagnetic force between the ith stator and the jth rotor teeth. Figure 4.6 shows the air-gap permeance and its space derivative between the ith stator and the jth rotor teeth for unskewed and skewed conditions. It is seen that by skewing, the magnitude of the air-gap permeance decreases. However, its space derivative is smoother [1].

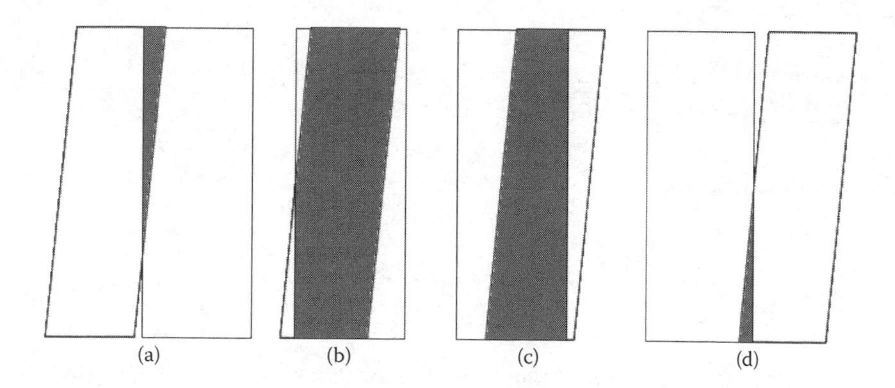

FIGURE 4.5
Different shapes of bases for positions of a skewed rotor tooth with respect to a stator tooth. Shaded area is the basis of flux tube as the rotor tooth moves.

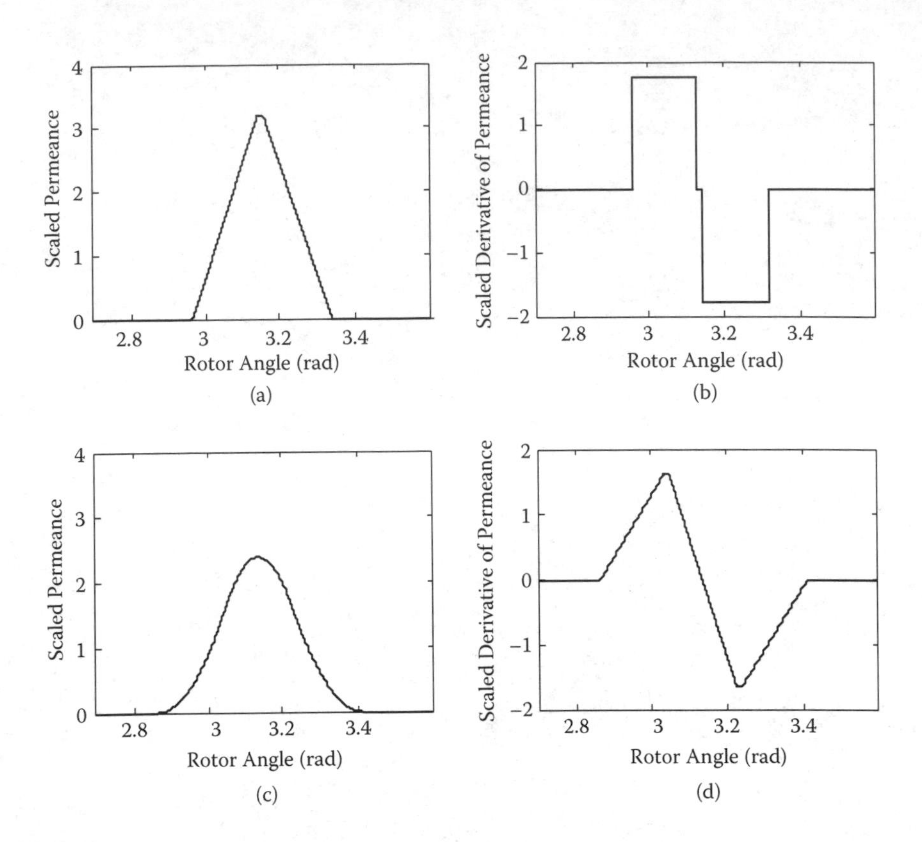

FIGURE 4.6
Air-gap permeance for (a) unskewed rotor and (b) its derivative; (c) skewed rotor and (d) its derivative.

4.2 Indirect Application of Magnetic Equivalent Circuit for Analysis of Salient Pole Synchronous Machines

The well-known equations of synchronous or induction machines that relates the flux linkages, stator and rotor currents, and self and mutual inductances are

$$\begin{aligned} L_{ss}\, i_s + L_{sr}\, i_r &= \lambda_s \\ L_{rs}\, i_s + L_{rr} i_r &= \lambda_r \end{aligned} \tag{4.5}$$

Using the magnetic equivalent circuit of synchronous and induction machines [1], the machines inductances L_{ss}, L_{sr}, L_{rs}, and L_{rr} will be determined. Neglecting iron saturation and using the magnetic equivalent circuit of synchronous machines, the inductance coefficients of the machines are obtained. These inductances can be used for analysis and study of the machine behavior under healthy and faulty conditions.

4.2.1 Magnetic Equivalent Circuit of a Salient Pole Synchronous Machine

A part of magnetic equivalent circuit of a typical salient-pole synchronous machine [2] is shown in Figure 4.7.

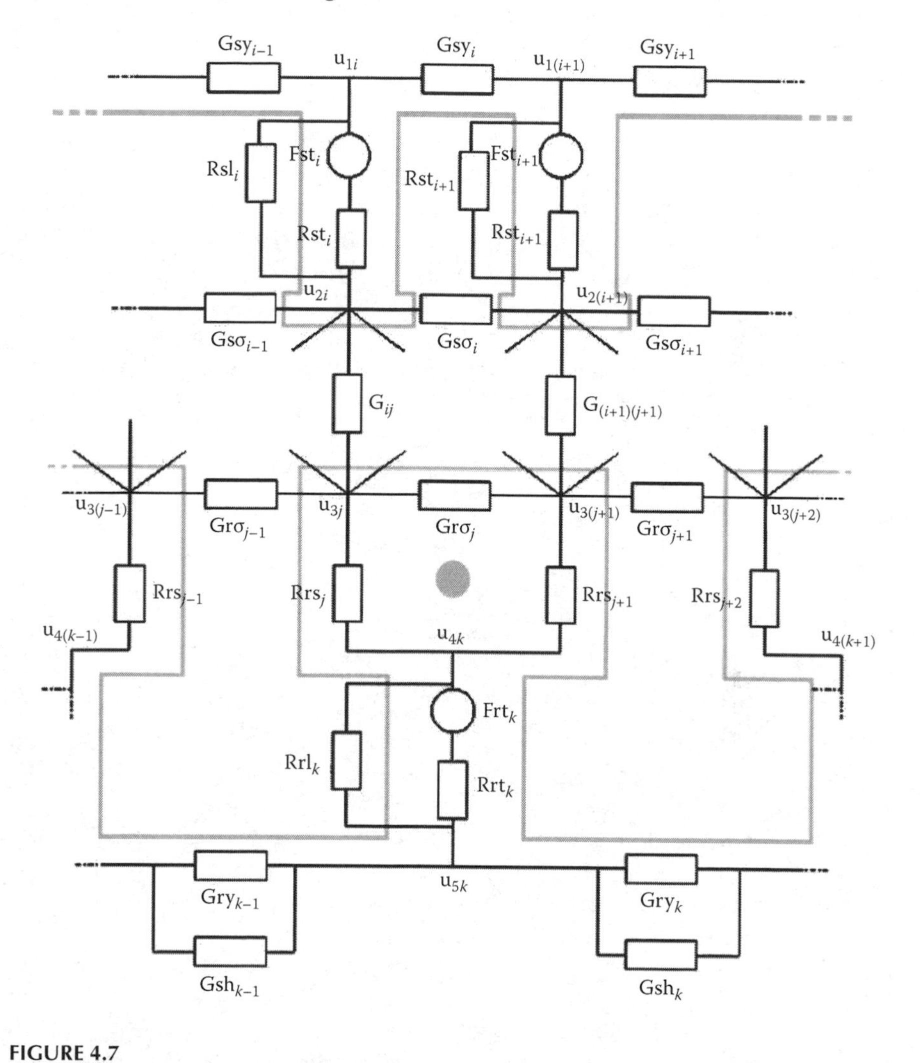

FIGURE 4.7
A part of the magnetic equivalent circuit of a salient pole synchronous machine, including all permeances and reluctances.

For the analysis the following assumptions are made:

1. Flux orientation from stator to rotor crosses the air-gap in radial direction.
2. Linear magnetic characteristic is considered.
3. For simplification no damper winding effect is considered.
4. Slot effects are considered.
5. Eddy current is neglected.

Each rotor pole core is divided into two parts. Damper windings are located in the outer part and field windings are wound around the inner part of each rotor pole. Here, damper windings are not considered. Therefore, only outer part of each rotor pole is considered having two segments represented by their reluctances. As shown R_{rsj} and R_{rsj+1} are the reluctances of these sections. R_{rt} is the rotor shank reluctance and R_{st} is the stator tooth reluctance. R_{sl} and R_{rl} are tooth leakage reluctances of stator and rotor, respectively. $G_{s\sigma}$ is the stator openings leakage permeance and G_{sy} is the stator yoke permeance. $G_{r\sigma}$ is the rotor openings leakage permeance and G_{ry} is the rotor yoke permeance. G_{sh} indicates the permeance of the rotor shaft, which is in parallel with the rotor yoke permeances.

There are five levels of MMF nodes defined as u_1, u_2, u_3, u_4, and u_5 vectors, which are used in writing the nodal equations of the machine. The number of rotor poles is p, number of sections on each rotor pole is m, and number of stator slots is n.

According to the magnetic equivalent circuit of Figure 4.7, the nodal equations are written as

$$A_{11}\, u_1 + A_{sl}\, u_2 = -\Phi_{st} \tag{4.6}$$

$$A_{sl}\, u_1 + A_{22}\, u_2 + A_{23}\, u_3 = \Phi_{st} \tag{4.7}$$

$$A_{32}\, u_2 + A_{33}\, u_3 + A_{34}\, u_4 = 0 \tag{4.8}$$

$$A_{43}\, u_3 + A_{44}\, u_4 + A_{rl}\, u_5 = \Phi_{rt} \tag{4.9}$$

$$A_{rl}\, u_4 + A_{55}\, u_5 = -\Phi_{rt} \tag{4.10}$$

The vectors Φ_{st} and Φ_{rt} contain the stator and rotor teeth fluxes. A_{11}, A_{22}, A_{33}, A_{44}, A_{55}, A_{23}, A_{32}, A_{34} and A_{43}, are the node permeance matrices, which are constructed according to the magnetic equivalent circuit theory concepts. The first two matrices A_{11} and A_{22} are $n \times n$, A_{44} and A_{55} are $p \times p$ matrices, and A_{33} is an $m \times m$ matrix. A_{sl} and A_{rl} are $n \times n$ and $p \times p$ matrices, respectively, and contain the leakage permeances of stator teeth and rotor shank. A_{23} is an $n \times p$ matrix containing the air-gap permeances, G_{ij}. A_{32} is the transpose of A_{23}. A_{34} is an $m \times n$ matrix containing the rotor-sections permeances, and its transpose is A_{43}.

The two following equations relate the magnetic potentials of yoke parts and teeth in the stator and the rotor sides, respectively,

$$u_2 = u_1 - R_{st}\Phi_{st} + F_{st} \tag{4.11}$$

$$u_4 = u_5 - R_{rt}\Phi_{rt} + F_{rt} \tag{4.12}$$

where F_{st} and F_{rt} are the MMF vectors of the stator and the rotor. F_{st} is related to stator current vector by

$$F_{st} = W_s i_s \tag{4.13}$$

and F_{rt} is related to the rotor current vector by

$$F_{rt} = W_r i_r \tag{4.14}$$

W_s is $n \times 3$ and W_r is $p \times p$ matrices. W_s describes the stator winding configuration of stator windings of the machine. W_s is called the MMF transform matrix. It has an important role in the calculation of machine inductance coefficients. Each column of W_s corresponds to the winding function (winding distribution in slots) of one of the stator phases. W_r is a diagonal matrix describing the winding turns per rotor poles.

4.2.2 Inductance Relations of a Salient Pole Synchronous Machine

Using Equation (4.6) u_2 is calculated as the following:

$$u_2 = -A_{sl}^{-1} A_{11} u_1 - A_{sl}^{-1} \Phi_{st} \tag{4.15}$$

Combining Equations (4.11), (4.13), and (4.15) leads to

$$-A_{sl}^{-1} A_{11} u_1 - A_{sl}^{-1} \Phi_{st} - u_1 + R_{st} \Phi_{st} = W_s i_s$$

$$(-A_{sl}^{-1} A_{11} - I_{n\times n})u_1 = -(A_{sl}^{-1} + R_{st})\Phi_{st} + W_s i_s$$

By defining $B_{n\times n}$ and $N_{n\times n}$ as

$$B_{n\times n} = (-A_{sl}^{-1} A_{11} - I_{n\times n})$$

$$N_{n\times n} = -B^{-1}(-A_{sl}^{-1} + R_{st})$$

u_1 and u_2 are related to the stator currents and stator teeth fluxes as

$$u_1 = -B^{-1}(-A_{sl}^{-1} + R_{st})\Phi_{st} + B^{-1}W_s i_s$$
$$\Rightarrow u_1 = N\Phi_{st} + B^{-1}\, W_s i_s \tag{4.16}$$

$$u_2 = [-B^{-1}(-A_{sl}^{-1} + R_{st}) - R_{st}]\Phi_{st} + (B^{-1}W_s + W_s)i_s \tag{4.17}$$

By defining C and F by

$$C_{n\times n} = -B^{-1}(-A_{sl}^{-1} + R_{st}) - R_{st}$$
$$F_{n\times 3} = (B^{-1}W_s + W_s)$$

Equation (4.17) simplifies to

$$u_2 = C\Phi_{st} + Fi_s \tag{4.18}$$

By the same procedure for rotor Equations (4.9) and (4.10), starting from Equation (4.10), and by defining the following matrices:

$$D_{m\times m} = (-A_{rl}^{-1}\, A_{55} - I_{m\times m})$$
$$M_{m\times m} = -D^{-1}(-A_{rl}^{-1} + R_{rt})$$
$$E_{m\times m} = -D^{-1}(-A_{rl}^{-1} + R_{rt}) - R_{rt}$$
$$G_{m\times m} = (D^{-1}\, W_r + W_r)$$

u_4 and u_5 are related to the rotor currents and rotor teeth fluxes as the following:

$$u_4 = E\,\Phi_{rt} + Gi_r \tag{4.19}$$
$$u_5 = M\,\Phi_{rt} + D^{-1}\, W_r i_r \tag{4.20}$$

By inserting u_2 from Equation (4.18) and u_4 from Equation (4.19) into Equation (4.8), u_3 is obtained as the following:

$$u_3 = -A_{33}^{-1}(A_{32}\, C\Phi_{st} + A_{32}\, Fi_s + A_{34}E\Phi_{rt} + A_{34}\, Gi_r$$
$$\Rightarrow u_3 = (-A_{33}^{-1}\, A_{32}\, C)\Phi_{st} + (-A_{33}^{-1}A_{32}F)i_s$$
$$+ (-A_{33}^{-1}\, A_{34}\, E)\Phi_{rt} + (-A_{33}^{-1}A_{34}G)i_r$$

where by defining the following matrices

$$H_{m\times m} = (-A_{33}^{-1}\, A_{32}\, C)$$
$$J_{m\times 3} = (-A_{33}^{-1}\, A_{32}\, F)$$
$$K_{m\times m} = (-A_{33}^{-1}\, A_{34}\, E)$$
$$L_{m\times m} = (-A_{33}^{-1}\, A_{34}\, G)$$

u_3 is related to the stator and rotor currents and stator teeth fluxes as the following:

$$u_3 = H\Phi_{st} + Ji_s + K\Phi_{rt} + Li_r \tag{4.21}$$

Replacing Equations (4.16), (4.18), and (4.21) into Equation (4.7) gives

$$A_{sl}\, N\,\Phi_{st} + A_{sl}\, B^{-1}W_s i_s + A_{22}\, C\Phi_{st} + A_{22}\, Fi_s$$

$$+A_{23}\, H\,\Phi_{st} + A_{23}\, Ji_s + A_{23}\, K\Phi_{rt} + A_{23}\, Li_r = \Phi_{st}$$

Rearranging the terms in the preceding relation leads to

$$[I_{n\times n} - A_{sl}\, N - A_{22}\, C - A_{23}\, H]\Phi_{st} - A_{23}\, K\,\Phi_{rt}$$

$$= [A_{sl}\, B^{-1}\, W_s + A_{22}\, F + A_{23} J]i_s + A_{23}\, Li_r$$

where by defining the following matrices

$$P_{n\times n} = I_{n\times n} - A_{sl}\, N - A_{22}\, C - A_{23} H$$

$$Q_{n\times 3} = A_{sl}\, B^{-1} W_s + A_{22} F + A_{23} J$$

leads to

$$P\Phi_{st} - A_{23}\, K\Phi_{rt} = Qi_s + A_{23}\, Li_r$$

$$\Rightarrow \Phi_{st} - P^{-1} A_{23} K\Phi_{rt} = P^{-1} Qi_s + P^{-1}\, A_{23}\, Li_r$$

Now by multiplying both sides of the preceding relation by W_s^T

$$W_s^T\,\Phi_{st} - W_s^T\, P^{-1}\, A_{23}\, K\Phi_{rt} = W_s^T\, P^{-1}\, Qi_s + W_s^T\, P^{-1}\, A_{23}\, Li_r$$

and defining the left-hand side of the result as stator flux linkage λ_s results in

$$\lambda_s = W_s^T\, P^{-1}\, Qi_s + W_s^T\, P^{-1}\, A_{23}\, Li_r \tag{4.22}$$

Now using the same procedure and replacing Equations (4.19), (4.20), and (4.21) into Equation (4.9) gives

$$A_{43}\, H\Phi_{st} + A_{43}\, J\, i_s + A_{43}\, K\Phi_{rt} + A_{43} Li_r + A_{44} E\Phi_{rt}$$

$$+ A_{44}\, Gi_r + A_{ri}\, M\Phi_{rt} + A_{rl}\, D^{-1}\, W_r i_r = \Phi_{rt}$$

$$(A_{43} H)\Phi_{st} + [A_{43} K + A_{44} E + A_{rl} M - I_{m\times m}]\Phi_{rt}$$

$$= (-A_{43} J)i_s + [-A_{43} L - A_{44} G - A_{rl} D^{-1} W_r]i_r$$

where by defining the following matrices

$$R_{m\times m} = A_{43}K + A_{44}E + A_{rl}\,M - I_{m\times m}$$

$$S_{m\times 3} = -A_{43}L - A_{44}\,G - A_{rl}\,D^{-1}\,W_r$$

it simplifies to

$$(A_{43}H)\Phi_{st} + R\Phi_{rt} = (-A_{43}J)i_s + Si_r$$

$$\Rightarrow R^{-1}(A_{43}H)\Phi_{st} + \Phi_{rt} = R^{-1}(-A_{43}J)i_s + R^{-1}Si_r$$

Now by multiplying both sides of the previous relation by W_r^T

$$W_r^T\,R^{-1}(A_{43}H)\Phi_{st} + W_r^T\Phi_{rt}$$

$$= W_r^T\,R^{-1}(-A_{43}J)i_s + W_r^T R^{-1}Si_r$$

and defining the left-hand side of the result as rotor flux linkage λ_r results in

$$\lambda_r = W_r^T\,R^{-1}(-A_{43}J)i_s + W_r^T R^{-1}Si_r \tag{4.23}$$

Now by comparing Equation (4.22) and Equation (4.23) with Equation (4.5) the following equations are obtained for the calculation of inductance coefficients of a salient pole synchronous machine

$$L_{ss} = W_s^T\,P^{-1}Q \tag{4.24}$$

$$L_{sr} = W_s^T\,P^{-1}A_{23}L \tag{4.25}$$

$$L_{rs} = W_r^T\,R^{-1}(-A_{43}J) \tag{4.26}$$

$$L_{rr} = W_r^T\,R^{-1}S \tag{4.27}$$

It should be noted that due to the inclusion of all stator and rotor reluctances it is possible to study the effect of the magnetic properties on all inductances by this model. Here only linear magnetic curve is studied.

4.2.3 Calculation of Inductances for a Salient Pole Synchronous Machine

Using Equations (4.24) to (4.27) all inductance coefficients are calculated for a 9 kVA, three-phase, four-pole salient pole synchronous machine as

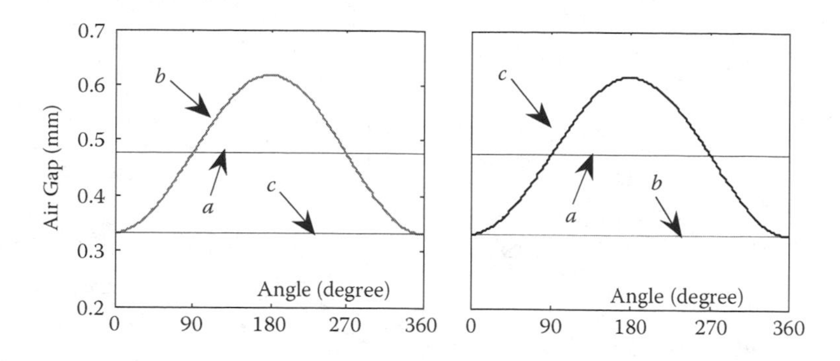

FIGURE 4.8
Air-gap curves for (a) symmetric, (b) 30% static eccentric, and (c) 30% dynamic eccentric machine from the viewpoint of a fixed point on the rotor (left) and a fixed point on the stator (right).

an example. Effects of conventional air-gap asymmetries such as static, dynamic, and mixed eccentricities on machine inductances are studied. These inductances are known as static eccentricity (SE), dynamic eccentricity (DE), and mixed eccentricity (ME), respectively. It is assumed that the reference points for both SE and DE are the same. This reference point is at zero angle. The middle point of the first stator tooth and first rotor pole are set at zero angles.

The left panel of Figure 4.8 shows air-gap variations seen from the reference point on the rotor side for eccentric and non-eccentric conditions. The right panel of Figure 4.8 shows these variations seen from the reference point on the stator side.

Figure 4.9 depicts the effective air-gap width for the simulated machine. Stator slot openings, rotor saliency effects, and the shape of rotor poles are considered. All these effects are included in the simulation. Maximum width corresponds with the rotor openings. Exact modeling of the air-gap is the

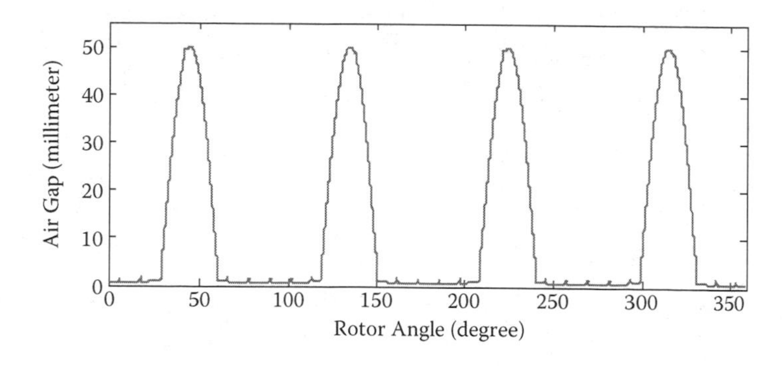

FIGURE 4.9
Effective air-gap width for the simulated machine for healthy condition.

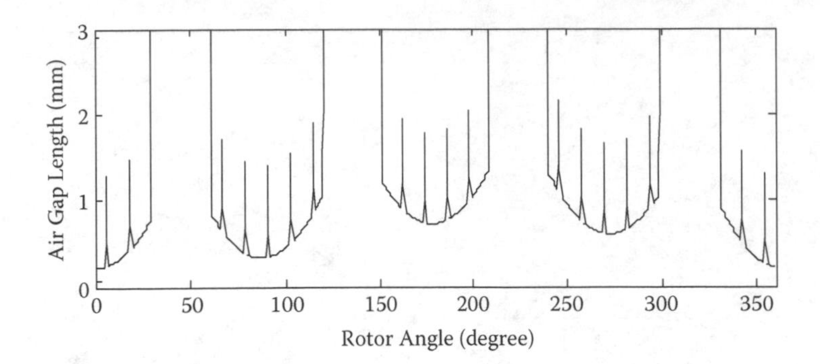

FIGURE 4.10
Zoomed effective air-gap width for the simulated machine with mixed air-gap eccentricity.

most important task for calculating machine inductances. Figure 4.10 shows the air-gap variation for mixed air-gap eccentricity. There are 30% static and 30% dynamic eccentricities. The effect of the rotor pole shape can be seen in this figure [2].

In the following we calculate machine inductances for symmetric and non-symmetric air-gap conditions. When there is any type of individual static or dynamic eccentricities, the amount of each eccentricity is considered to be 30%. Therefore, in mixed air-gap eccentricity there exists 30% static and 30% dynamic eccentricities.

Figure 4.11 shows the stator self-inductances for a symmetric air-gap and for a mixed air-gap eccentricity. Dashed curve is related to healthy machine and solid curve is related to faulty machine. Whereas the maximum magnitude of stator self-inductance in this figure mostly depends on the air-gap width under the rotor pole arc, the minimum magnitude depends on the rotor depth between the rotor poles. It can be seen that under mixed air-gap eccentricity, maximum points have more variation with respect to the minimum points. The reason is that the percentage variation of air-gap width under rotor pole arc is much more

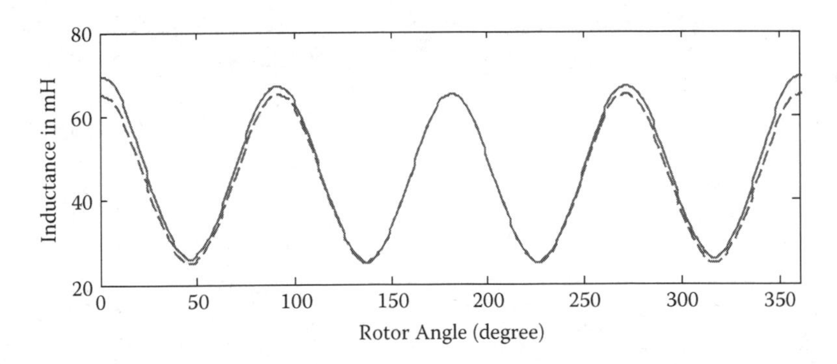

FIGURE 4.11
Stator self-inductances for symmetric and ME (SE: 30% and DE: 30%) conditions.

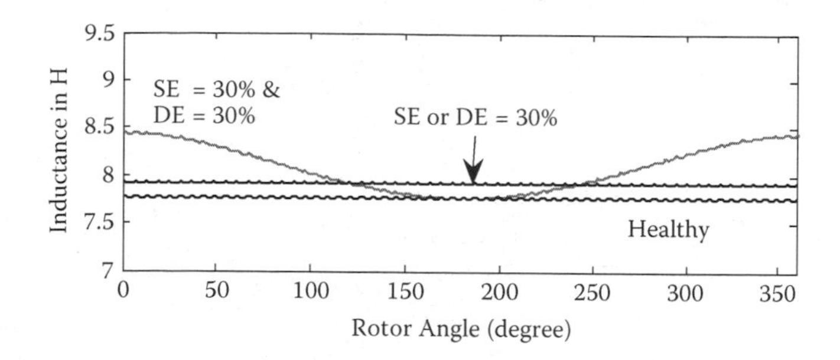

FIGURE 4.12
Self-inductance of one of the rotor windings for healthy (SE: 30%, DE: 30%) and ME (SE: 30% and DE: 30%) conditions.

than the percentage variation of rotor depth between rotor poles. In the case of individual static or dynamic eccentricities no significant variation is observed.

Figure 4.12 shows the rotor self-inductances for a symmetric air-gap, for an individual static or dynamic air-gap eccentricity, and for a mixed air-gap eccentricity. The effects on individual static eccentricity and dynamic eccentricity are the same. Both just increase the rotor self-inductance. However, in mixed air-gap eccentricity the rotor self-inductance varies with the rotor position. It is concluded that this may be a suitable measure for the existence of mixed air-gap eccentricity in synchronous machine. The effect of the stator slot openings is apparent in the rotor self-inductance.

Figure 4.13 shows the stator mutual inductances for a symmetric air-gap and for a mixed air-gap eccentricity. A dashed curve is related to a healthy machine and a solid curve is related to a faulty machine. As expected, mutual inductances have negative values. In this figure, the most negative points of the stator mutual inductance are mostly dependent on the air-gap width under rotor pole arc and the least negative points are dependent on the rotor depth between the rotor poles.

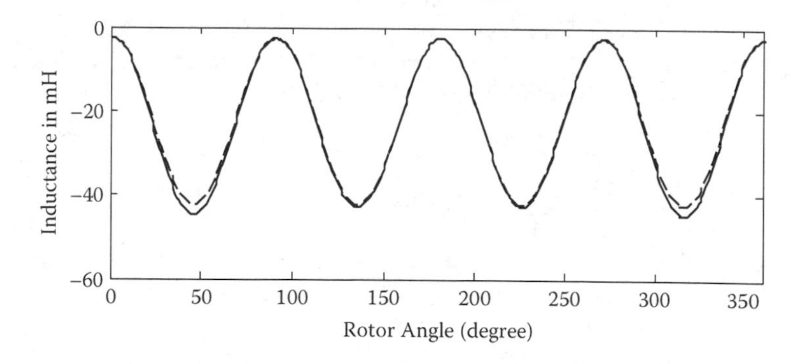

FIGURE 4.13
Stator mutual inductances compared for healthy and ME (SE: 30% and DE: 30%) conditions.

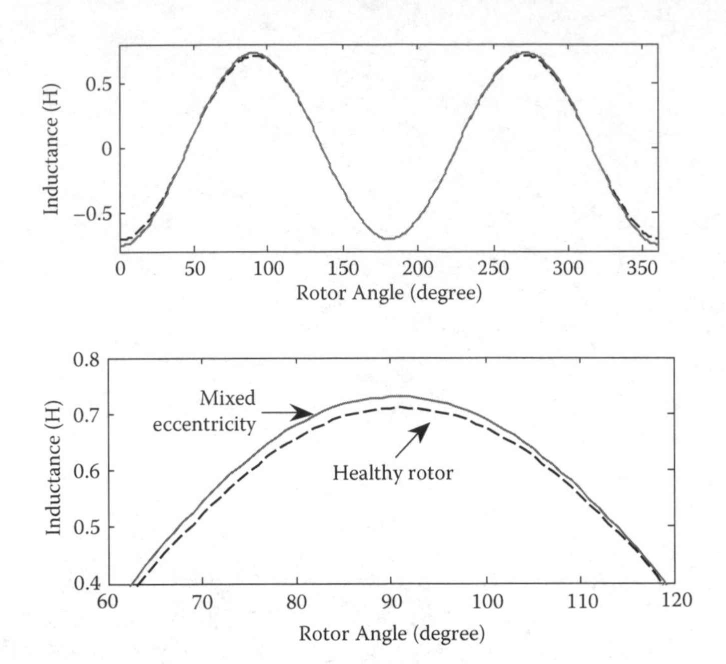

FIGURE 4.14
Mutual inductances between one of the stator phases and rotor winding for the healthy machine and faulty machine with mixed air-gap eccentricity (SE: 30% and DE: 30%).

It can be seen that under mixed air-gap eccentricity the most negative points have more variation with respect to the least negative points. The reason is the same as for the stator self-inductance. Again the curves of stator mutual inductances for static and dynamic eccentricities are approximately the same and are very close to the healthy case.

Figure 4.14 represents the mutual inductance curves of one of the stator phases and the rotor winding for a symmetric air-gap and for a mixed air-gap eccentricity. Dashed curve is related to healthy machine and solid curve is related to faulty machine. As expected, the mutual inductance has both positive and negative values. To show the effects of air-gap fault on this inductance a zoomed part of top plot is also shown. It is seen that the effect of the same value of mixed eccentricity on stator to rotor mutual inductance is less than the effect on previous inductances.

Skewing of stator or rotor slots is a technical manufacturing method for better machine performance. With skewed stator or rotor slots, machine inductances vary more smoothly. In other words, slot opening degrades the mutual inductances. Skew can recover it again [4]. In the derived equation for calculating machine inductances, skew effect can be included.

In Figure 4.15, the plots of stator to rotor mutual inductances for a skewed and for an unskewed stator slots are shown. Comparison of the plots in Figure 4.15 shows how a skewed slot affects the shape of these mutual

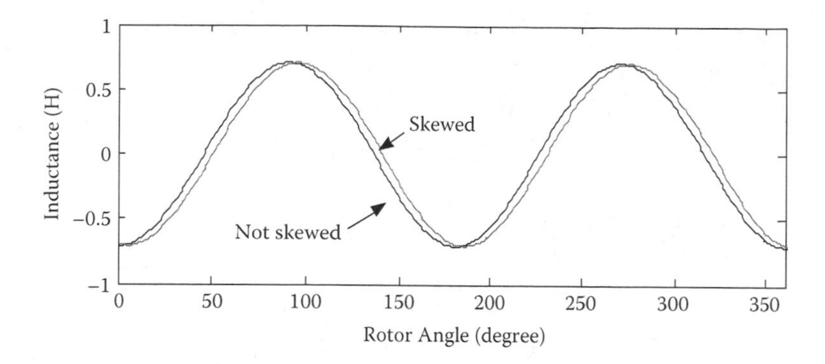

FIGURE 4.15
Comparison of mutual inductances between one of the stator phases and the rotor winding for skewed and unskewed slots.

inductances. Magnitude of mutual inductances in the skewed case is a little less with respect to unskewed inductances. However, the ripples due to slot openings are suppressed in this plot. To further study the effect of skew, the derivative of mutual inductance is calculated and plotted in Figure 4.16. Effect of stator slot openings in an unskewed machine is clear. For a skewed slot the derivative of the mutual inductance variation is smooth.

4.2.4 Experimental Measurement of Inductances of a Salient Pole Synchronous Machine

The experimental investigation was carried out to verify the theoretical and simulation findings. The same synchronous machine that is used for simulation is used in experiments. Design data for this machine are given in Appendix A. This machine is coupled to a three-phase induction motor.

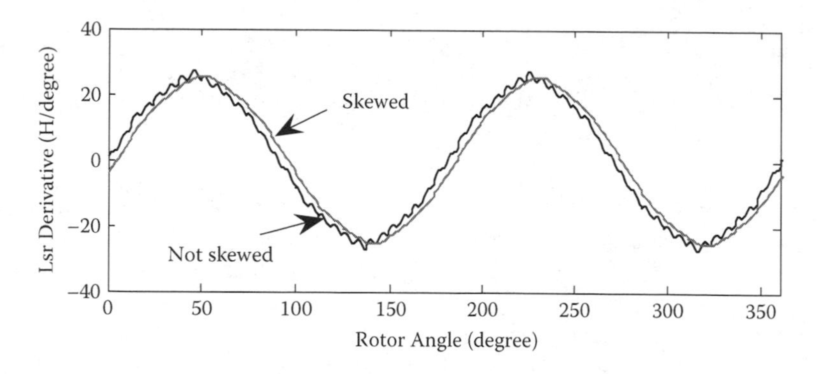

FIGURE 4.16
Derivative of mutual inductances for skewed and unskewed conditions.

To measure the mutual inductance between windings p and q, L_{pq}, a direct current voltage is applied to the second winding. The synchronous machine is run by an induction motor. The current in the second winding is measured and saved by a digital oscilloscope. The induced voltage in the first winding is also measured and saved. Faraday's law implies that

$$e = \frac{d\lambda}{dt} = \frac{d(L_{pq}i)}{dt} \tag{4.28}$$

Integrating both sides of the preceding equation leads to

$$\lambda = \int e.dt = \int d(L_{pq}i) = L_{pq}i \tag{4.29}$$

Then

$$L_{pq} = \frac{\int e.dt}{i} \tag{4.30}$$

To measure the self-inductance of windings k, L_{kk}, a direct voltage is applied to this winding. The synchronous machine is run by an induction motor. The current in the winding and the applied voltage are measured and saved.

The terminal voltage equation for the supplied winding is given by

$$v = R_k i + e = R_k i + \frac{d(L_{kk}i)}{dt} \tag{4.31}$$

Manipulating the preceding equation and integrating both sides leads to

$$\int (v - R_k i).dt = \int e.dt = \int d(L_{kk}i) = L_{kk}i \tag{4.32}$$

Then

$$L_{kk} = \frac{\int (v - R_k i)dt}{i} \tag{4.33}$$

It should be noted that in the calculation of self-inductance it is necessary to measure winding resistance.

According to Equations (4.28) to (4.33), four inductances of synchronous machine are measured and plotted in Figures 4.17 to 4.20. It is necessary to note that the given experimental plots have been filtered. To see the effect of filtering, the unfiltered plot of stator self-inductance is shown in the top of Figure 4.17.

In the first view, comparison of inductance plots extracted from the model and the derived equations with inductance plots obtained from the

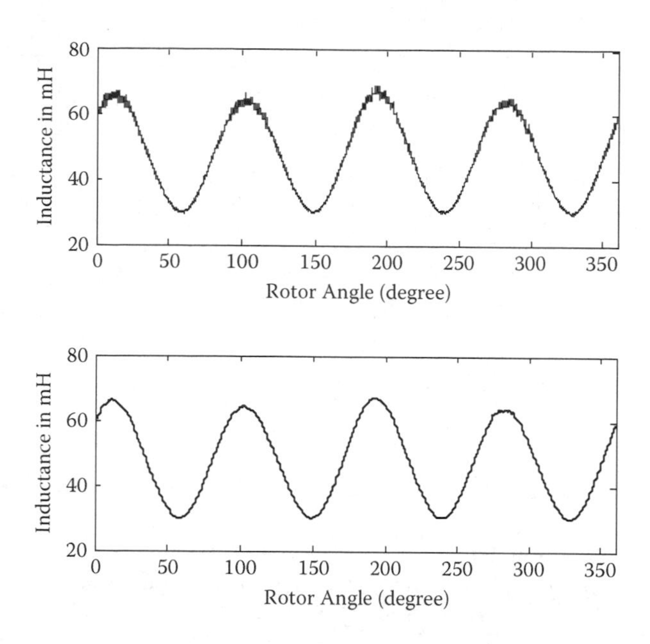

FIGURE 4.17
Stator self inductances resulted from the experiment (a) before filtering and(b) after filtering.

experiment show very good agreements. More investigation into experimental plots shows that in the stator self-inductance, the magnitude of consecutive peaks is different. This may be due to some kind of rotor misalignment or nonlinear behavior of the machine core. In the two-dimensional model used in this textbook these conditions are not included. However, the method can be developed for the analysis of three-dimensional conditions.

It is seen that there are some ripples in the rotor self inductance in Figure 4.18. It may be due to rotor misalignment or nonlinear behavior of the machine core.

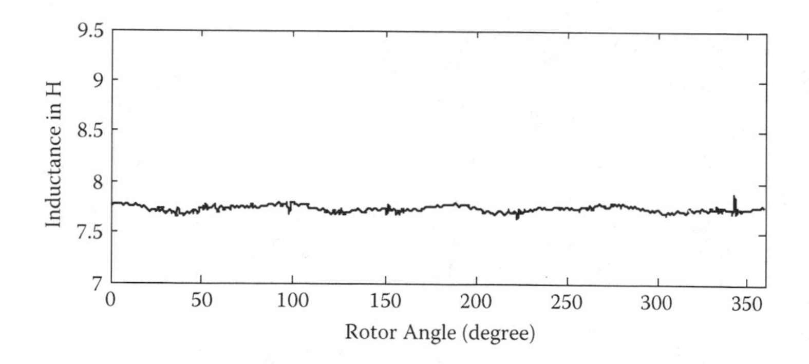

FIGURE 4.18
Rotor self-inductance curve.

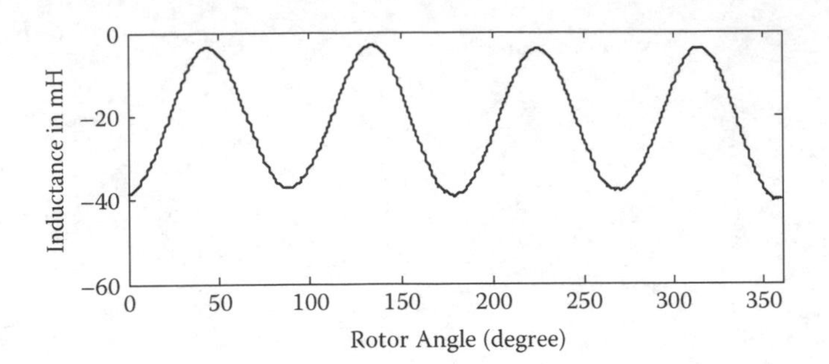

FIGURE 4.19
Mutual inductance between two stator phases.

4.3 Indirect Application of Magnetic Equivalent Circuit for Analysis of Induction Machines

By using a simplified magnetic equivalent circuit of an induction machine we obtain the inductance coefficients of the machines [3,4]. These inductances can be used for analysis and study of the machine behavior in healthy and under faulty conditions.

4.3.1 A Simplified Magnetic Equivalent Circuit of Induction Machines

In Figure 4.21, a part of a magnetic equivalent circuit of an induction machine is shown [3,4]. u_1, u_2, u_3, and u_4 are the vectors of magnetic node potential in stator back iron, stator teeth, rotor teeth, and rotor back iron, respectively.

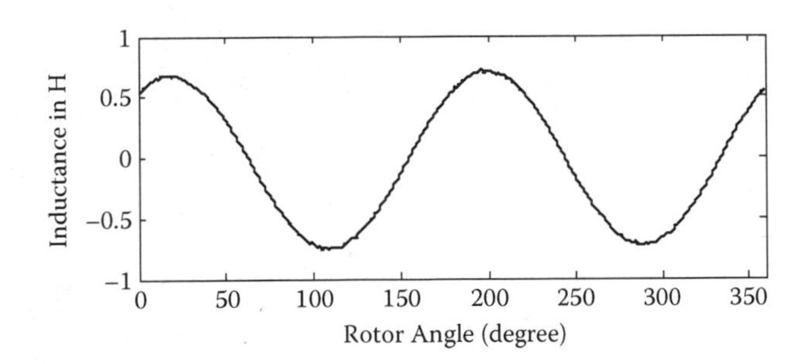

FIGURE 4.20
Mutual inductance between phase *a* of stator and rotor winding.

The node potential equations for the network of Figure (4.21) are given as

$$A_{11}\,u_1 = -\Phi_{st} \tag{4.34}$$

$$A_{22}u_2 + A_{23}u_3 = \Phi_{st} \tag{4.35}$$

$$A_{32}u_2 + A_{33}u_3 = \Phi_{rt} \tag{4.36}$$

$$A_{44}u_4 = -\Phi_{rt} \tag{4.37}$$

$$u_2 = u_1 - R_{st}\Phi_{st} + F_{st} \tag{4.38}$$

$$u_3 = u_4 - R_{rt}\Phi_{rt} + F_{rt} \tag{4.39}$$

where Φ_{st} and Φ_{rt} are the vectors of stator and rotor teeth fluxes. R_{st} and R_{rt} are stator and rotor teeth reluctance matrices. F_{st} and F_{rt} are the vectors of MMF sources in the stator side and rotor side, respectively. A_{11}, A_{22}, A_{23}, A_{32}, A_{33}, and A_{44} are the node permeance matrices and are given in Appendix B. The element of A_{11} and A_{44} matrices depends only on stator and rotor back iron segment permeances, respectively. In Figure 4.21, these permeances

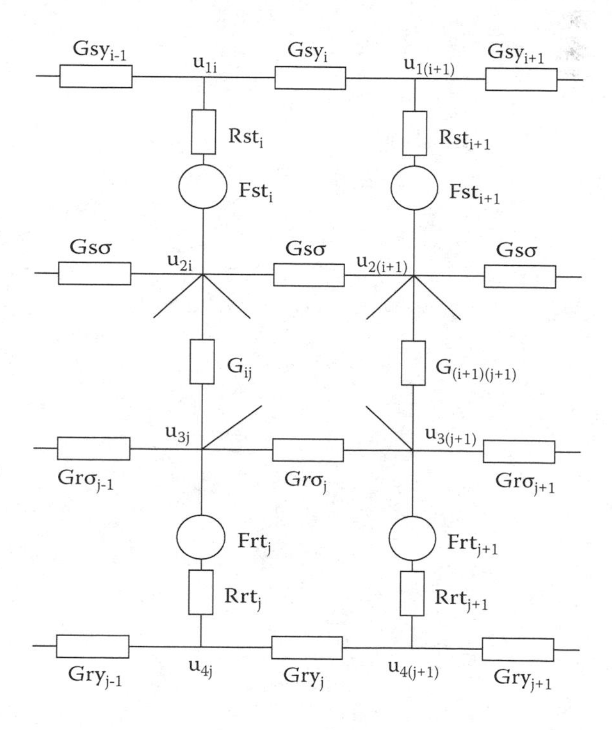

FIGURE 4.21
A part of magnetic equivalent circuit of an induction machine.

are shown by $G_{sy,i}$ and $G_{ry,j}$. The elements of A_{22}, A_{23}, A_{32}, and A_{33} depend on stator slot, rotor slot, and air-gap permeances. $G_{s\sigma}$ is the stator slot openings leakage permeance and is constant. $G_{r\sigma,j}$ is the rotor slot leakage permeance of slot j, and for closed slot, due to saturation effect, it is a nonlinear permeance. G_{ij} is the air-gap permeance between stator tooth i and rotor tooth j. G_{ij} is the most important parameter in the magnetic equivalent circuit modeling. Derivative of the air-gap permeance G_{ij} with respect to the rotor angle when multiplied by the square of the MMF drop over the same permeance gives the value of electromagnetic force between them.

F_{st} and F_{rt} vectors are related to the stator phase currents and rotor mesh currents through the following equations

$$F_{st} = W_s i_s \tag{4.40}$$

$$F_{rt} = W_r i_r \tag{4.41}$$

where W_s is generated using the same approaches as for the synchronous machine. W_r is an identity matrix and its size depends on the number of independent rotor mesh currents i_r.

4.3.2 Inductance Relations of Induction Machines

Equation (4.34) and Equation (4.37) are written for the back iron portions of the stator and rotor, respectively. Figure 4.22 shows a portion of a typical rotor lamination of an induction machine. The geometric shape of back iron parts in stator and rotor are such that they have a large cross-sectional area and nearly short length with respect to the teeth segments of stator and rotor. As a result, the MMF drops in these back iron parts are generally several times smaller than MMF drops on teeth segments. It should be noted that in the

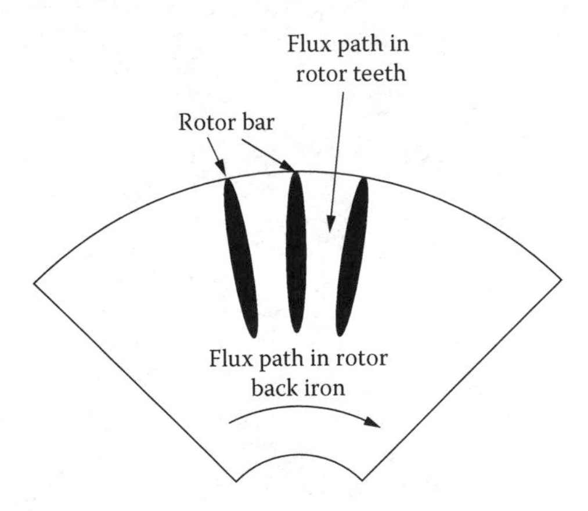

FIGURE 4.22
A portion of a typical rotor lamination of an induction machine.

MEC model, it is assumed that the directions of fluxes in a tooth and in a back iron segment are perpendicular to each other. Although back iron segments reluctances have certainly some effects on the values of machine inductances, simulation results show actually that neglecting the MMF drops in back iron segments has a very small effect on machine inductance coefficients. On the other hand, it is possible to change the values of teeth reluctances by some percentage to compensate for the removal of back iron reluctances.

Neglecting the back iron reluctances in stator leads to equality of u_1 elements and neglecting the back iron reluctances in rotor leads to equality of u_4 elements. On the other hand, due to the fact that

$$\sum_{i=1}^{n_s} \Phi_{st_i} = 0 \tag{4.42}$$

Therefore, $u_1 = 0$ and

$$u_2 = -R_{st}\Phi_{st} + W_s i_s \tag{4.43}$$

and

$$\sum_{j=1}^{n_r} \Phi_{rt_j} = 0 \tag{4.44}$$

Hence, $u_4 = 0$ and

$$u_3 = -R_{rt}\Phi_{rt} + i_r \tag{4.45}$$

The results of such assumptions lead to removal of Equations (4.34) and (4.37) from the system of algebraic equations of MEC model of induction machine shown by Equations (4.34) to (4.41).

By substituting Equations (4.38), (4.39), (4.43), and (4.45) in Equations (4.35) and (4.36), and then rearranging parameters, we have

$$A_{22}W_s i_s + A_{23} i_r = (I_{ns\times ns} + A_{22}R_{st})\Phi_{st} + A_{23}R_{rt}\Phi_{rt} \tag{4.46}$$

$$A_{32}W_s i_s + A_{33} i_r = (I_{nr\times nr} + A_{33}R_{rt})\Phi_{rt} + A_{32}R_{st}\Phi_{st} \tag{4.47}$$

By introducing matrices C and D as follows:

$$C = (I_{ns\times ns} + A_{22}R_{st})^{-1} \tag{4.48}$$

$$D = (I_{nr\times nr} + A_{33}R_{rt})^{-1} \tag{4.49}$$

and by further simplification the following equations will be obtained

$$C \cdot A_{22} W_s i_s + C \cdot A_{23} i_r = \Phi_{st} + C \cdot A_{23} R_{rt} \Phi_{rt} \tag{4.50}$$

$$D \cdot A_{32} W_s i_s + D \cdot A_{33} i_r = \Phi_{rt} + D \cdot A_{32} R_{st} \Phi_{st} \tag{4.51}$$

Multiplying both sides of Equation (4.50) by W_s^T and defining the right-hand side of the result as λ_s and also defining the right-hand side of Equation (4.51) as λ_r, yield

$$W_S^T CA_{22} W_s i_s + W_s^T CA_{23} i_r = \lambda_s \tag{4.52}$$

$$D \cdot A_{32} W_s i_s + D \cdot A_{33} i_r = \lambda_r \tag{4.53}$$

Comparing these equations with Equation (4.5) results in

$$L_{ss} = W_s^T CA_{22} W_s \tag{4.54}$$

$$L_{sr} = W_s^T CA_{23} \tag{4.55}$$

$$L_{rs} = DA_{32} W_s \tag{4.56}$$

$$L_{rr} = DA_{33} \tag{4.57}$$

In the preceding equations, the effects of all space harmonics, rotor skew, leakage path reluctances, and slot openings are taken into account for the calculation of inductance coefficients. Since there is no restriction concerning symmetry of stator windings, rotor bars, and air-gap length, this calculation may be applied in the study of asymmetrical effects and fault conditions on machine inductances. These fault conditions are short turns in stator windings and air-gap asymmetry such as the static and dynamic eccentricities. While short turns are reflected in the calculation of W_s, the air-gap asymmetries change the air-gap permeances and are included in matrices A_{22}, A_{23}, A_{32}, and A_{33}. It should be noted that due to the inclusion of stator and rotor teeth reluctances, it is possible to study the effect of the magnetic property of different cores on machine inductances by this model. Therefore, this model may be applied in the design of induction machines more efficiently.

4.3.3 Calculation of Inductance of an Induction Machine

Based on the equations derived for an induction machine, the inductance coefficients of a 3 hp, three-phase induction machine with the parameter given in Appendix A are calculated under different conditions. Table 4.1 and Figure 4.23 show some of the results of these calculations. The plots of mutual inductance between phase *a* of stator and one of the rotor loops, L_{ar}, for a healthy machine and for a machine with mixed air-gap eccentricity

TABLE 4.1

Calculation of Inductance Coefficient

Row Number	Stator Slot Opening	Reluctance Included/Skew Included	Machine Condition	L_{aa} (H)	L_{ab} (H)	Max. L_{ar} (mH)	Max. L_{rr} (μH)	Mean L_{rr} (μH)
1	0	N/Y	H	.3949	−.164	.13808	1.7047	1.7047
2	.03	N/Y	H	.3564	−.148	.1247	1.5385	1.5385
3	.06	N/Y	H	.3195	−.1327	.11238	1.3793	1.3793
4	.12	N/Y	H	.2506	−.1041	.092106	1.0817	1.0817
5	.15	N/Y	H	.2185	−.0908	.08512	.94335	.94335
6	0	N/N	H	.3949	−.164	.13808	1.7047	1.7047
7	.03	N/N	H	.3564	−.148	.1341	1.5386	1.5386
8	.06	N/N	H	.3195	−.1327	.13012	1.3795	1.3795
9	.03	Y/Y	H	.3509	−.145	(.12278)(.12301)*	1.3625	1.3625
10	.06	Y/Y	H	.3151	−.1309	(.11082)(.11100)*	1.3625	1.3625
11	.03	N/Y	SEC (20%)	.3645	−.151	.15567	1.941**	1.5883
12	.03	N/Y	DEC (20%)	.3645	−.151	.15549	1.9373	1.936

Note: SEC, static eccentricity condition; DEC, dynamic eccentricity condition; H, healthy machine.

* In these conditions, mutual inductance between phase *a* and rotor loop, L_{ar}, is different from L_{ra}.

** In SEC, L_{rr} is not constant and is dependent on the rotor position.

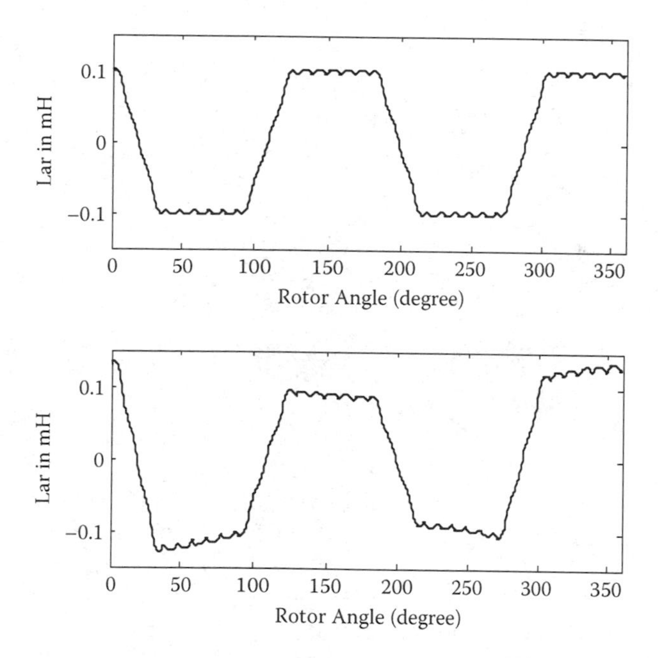

FIGURE 4.23

Plots of mutual inductance between phase *a* of stator and one of the rotor loops, L_{ar}, for a healthy machine (top) and for a machine with 5% dynamic eccentricity and 20% static eccentricity (bottom).

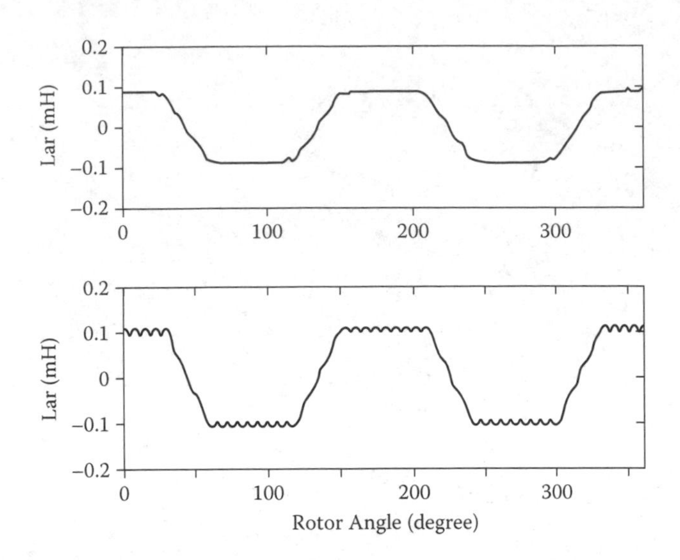

FIGURE 4.24
The plots of L_{ar} from FE calculation (top) and from the magnetic equivalent circuit (bottom) method.

are shown in Figure 4.23. There is 5% dynamic eccentricity and 20% static eccentricity in faulty condition. Stator and rotor slot openings are considered, and the rotor is skewed by one rotor pitch. It is seen that mixed air-gap eccentricity causes an asymmetrical mutual inductance between stator phase and rotor loop. The effects of slot openings are clear in both plots.

The plots of L_{ar} from FE calculation and from the magnetic equivalent circuit method are shown in Figure 4.24. Comparison of these plots shows that calculation of inductance coefficients by the MEC is in agreement with the FE method.

Comparison of rows 1 to 3 with rows 6 to 8 in Table 4.1 shows that as the stator slot opening increases, the magnitudes of mutual inductances between stator phases and rotor loops with skew and without skew effect have more differences. So for larger slot openings, increasing the skew causes more reduction in the magnitude of these mutual inductances.

Equations (4.54) and (4.57) show that teeth reluctances affect mutual inductances between stator phases and rotor loop. Also notice that L_{ar} is not the same as L_{ra}. However, comparing rows 2 and 3 with rows 9 and 10 in Table 4.1 shows that this effect is small.

Table 4.1 depicts the average of L_{aa} and L_{ab}, including slot width, rotor skew, and teeth reluctances effects for both healthy and eccentric rotors under different conditions. Again, it is seen that for both types of eccentricities there are considerable changes in stator inductances with respect to healthy condition, and this knowledge may be applied for the detection of these faults.

According to the inductance values shown in Table 4.1 and the plots of Figure 4.23, the following general results can be concluded:

1. By increasing the values of core reluctances, the values of inductance coefficients decrease. So, the effect of magnetic property of different cores can be studied in the design of induction motors.
2. Rotor skew affects mutual inductances between stator phases and rotor loops considerably.
3. By increasing the slot opening, the entire inductance coefficient values decrease.
4. Static eccentricity affects all inductance values.
5. Dynamic eccentricity affects all inductance values considerably.

4.4 Direct Application of Magnetic Equivalent Circuit Considering Nonlinear Magnetic Characteristic for Machine Analysis

Reluctances of stator and rotor tooth depend on the flux through them. Therefore, when nonlinear magnetic characteristics have to be considered in the analysis, it is not possible to calculate the inductance coefficients before the calculation of machine variables. In this case, the system of algebraic equations of the machine is nonlinear. Therefore an iterative procedure has to be employed to solve these equations [3,5]. In order to apply the solution procedure to the algebraic machine equation, some algebraic manipulations are required. These equations are repeated for convenience:

$$A_{22}u_2 + A_{23}u_3 = \Phi_{st} \tag{4.58}$$

$$A_{32}u_2 + A_{33}u_3 = \Phi_{rt} \tag{4.59}$$

$$u_2 = -R_{st}\Phi_{st} + W_s i_s \tag{4.60}$$

$$u_3 = -R_{rt}\Phi_{rt} + i_r \tag{4.61}$$

As the first step, Equations (4.60) and (4.61) are substituted in Equations (4.58) and (4.59). The results are arranged as

$$A_{22}W_s i_s + A_{23}i_r - (I_{n_s} + A_{22}R_{st})\Phi_{st} = R_{rt}\Phi_{rt} \tag{4.62}$$

$$A_{32}W_s i_s + A_{33}i_r - A_{32}R_{st}\Phi_{st} = (I_{n_r} + A_{33}R_{rt})\Phi_{rt} \tag{4.63}$$

Both sides of Equation (4.62) are multiplied by W_s^T and by defining $W_s^T.\Phi_{st}$ as the stator flux linkage λ_s the following equation will be obtained:

$$W_s^T A_{22} W_s i_s + W_s^T A_{23} i_r - W_s^T A_{22} R_{st} \Phi_{st} = \lambda_s + W_s^T A_{23} R_{rt} \Phi_{rt} \tag{4.64}$$

The vector M_0 is also defined to reflect the type of stator winding connection (delta, star, with or without a neutral connection). For a three-phase star connected without neutral connection, the sum of stator currents is zero. Hence,

$$i_a + i_b + i_c = 0$$

where by defining M_0 as the following

$$M_0 = [1 \quad 1 \quad 1]$$

we have

$$M_0 i_s = 0 \tag{4.65}$$

where

$$i_s = [i_a \quad i_b \quad i_c]^T$$

Equations (4.62) through (4.65) can be summarized in matrix notation as

$$\begin{bmatrix} M_0 & 0 & 0 \\ W_s^T A_{22} W_s & W_s^T A_{23} & -W_s^T A_{22} R_{st} \\ A_{22} W_s & A_{23} & -(I_{n_r} + A_{22} R_{st}) \\ A_{32} W_s & A_{33} & -A_{32} R_{st} \end{bmatrix} \begin{bmatrix} i_s \\ i_r \\ \Phi_{st} \end{bmatrix}$$

$$= \begin{bmatrix} 0 \\ \lambda_s + W_s^T A_{23} R_{rt} \Phi_{rt} \\ R_{rt} \Phi_{rt} \\ (I_{n_r} + A_{33} R_{rt}) \Phi_{rt} \end{bmatrix} \tag{4.66}$$

The preceding system of equations along with Equations (4.60) and (4.61) are used for the simulation of machines considering saturation characteristic. In each numerical step, after solving the differential equations of the machine, the right-hand side of the system of Equation (4.66) is calculated. However, the elements of the coefficient matrix in the left-hand side of Equation (4.66) depend on the unknown variables, currents, and teeth fluxes in the left-hand side of Equation (4.66). Therefore an iterative procedure for the solution of the system of Equation (4.66) has to be employed. In the following, one of the most popular methods that are based upon Newton's method is applied. In order to apply Newton's method to the system of Equation (4.66), using rows 2, 3, and 4 of the system of Equation (4.66), the functions F_1, F_2, and F_3 are defined as

$$F_1 = W_s^T A_{22} W_s i_s - W_s^T A_{22} R_{st} \Phi_{st} + W_s^T A_{23} (i_r - R_{rt} \Phi_{rt}) - \lambda_s \tag{4.67}$$

$$F_2 = -A_{22} W_s i_s + (I_{n_s} + A_{22} R_{st}) \Phi_{st} - A_{23} i_r + A_{23} R_{rt} \Phi_{rt} \tag{4.68}$$

$$F_3 = A_{32} W_s i_s - A_{32} R_{st} \Phi_{st} + A_{33} (i_r - R_{rt} \Phi_{rt}) - \Phi_{rt} \tag{4.69}$$

The following steps in the solution procedure are required:

1. Choose a set of values of unknown vectors i_s, i_r, and Φ_{st}.
2. Calculate the MMF drop on each nonlinear element of the magnetic equivalent circuit based upon values of the vectors from step 1.
3. Calculate the permeance of each nonlinear element knowing the MMF drop on it and its B-H characteristic.
4. Insert the computed permeances into matrices A_{22}, A_{33}, R_{st}, and R_{rt}.
5. Calculate the left hand side of the system of Equation (4.66).
6. Calculate the differences F_1, F_2, and F_3.
7. Choose a new set of unknowns from step 1 if the absolute values of differences calculated in step 6 are too big. If the error of step 6 is less than the maximum allowed, stop the process.

At the end of the iterative procedure the absolute values of the vectors of differences F_1, F_2, and F_3 have to be less than the specified error of computation. In other words, the solution of the system of Equation (4.66) is identical to the absolute minimum of Equations (4.67) to (4.69).

The Jacobian of the system of Equation (4.66) has to be defined. Therefore partial derivative of F_1, F_2, and F_3 vectors with respect to the unknowns need to be evaluated. This Jacobian is obtained as the following:

$$Jocobian = \begin{bmatrix} 1\ 1\ 1 & 0\cdots & 0 \\ \frac{\partial F_1}{\partial i_s} & \frac{\partial F_1}{\partial \Phi_{st}} & \frac{\partial F_1}{\partial i_r} \\ \frac{\partial F_2}{\partial i_s} & \frac{\partial F_2}{\partial \Phi_{st}} & \frac{\partial F_2}{\partial i_r} \\ \frac{\partial F_3}{\partial i_s} & \frac{\partial F_3}{\partial \Phi_{st}} & \frac{\partial F_3}{\partial i_r} \end{bmatrix} \tag{4.70}$$

where

$$\frac{\partial F_1}{\partial i_s} = W_s^T A_{22} W_s$$

$$\frac{\partial F_1}{\partial \Phi_{st}} = -W_s^T A_{22} R_{st}$$

$$\frac{\partial F_1}{\partial i_r} = W_s^T A_{23}$$

$$\frac{\partial F_2}{\partial i_r} = -A_{22} W_s$$

$$\frac{\partial F_2}{\partial \Phi_{st}} = I_{n_s} + A_{22} R_{st}$$

$$\frac{\partial F_2}{\partial i_r} = -A_{23}$$

$$\frac{\partial F_3}{\partial i_s} = A_{32} W_s$$

$$\frac{\partial F_3}{\partial \Phi_{st}} = -A_{32} R_{st}$$

$$\frac{\partial F_3}{\partial i_r} = A_{33}$$

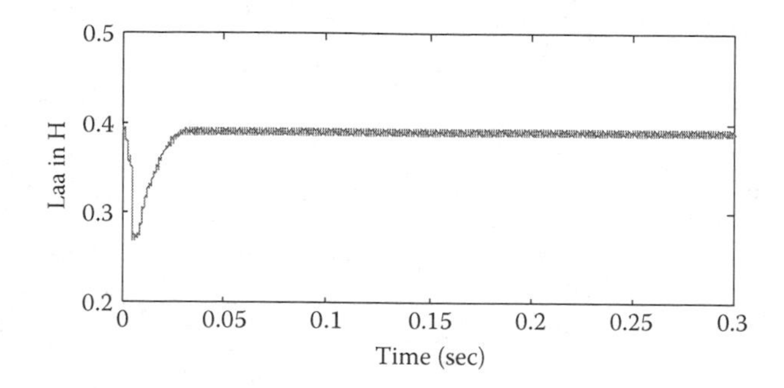

FIGURE 4.25
Variation of stator self-inductance versus time in free start of the motor.

As was mentioned before, the inductance coefficients depend on the stator and rotor fluxes, especially when there is deep saturation (refer to Equations 4.54 to 4.57), since they depend on the stator and rotor reluctances through matrices C and D). Figure 4.25 depicts this variation obtained from a free start of an induction motor in a complete simulation. This figure shows the variation of L_{aa} with respect to the rotor position in one rotor revolution that has been calculated in each step of simulation. As is shown, the average of L_{aa} in steady-state condition (0.39 H) is very close to the similar case (0.3949 H) calculated with a linear magnetic curve (Table 4.1). However, during the transient period the value of L_{aa} decreases considerably due to deep saturation since the stator currents are high. It should be noted for deep saturation in steady state, reduction of the average value of L_{aa} is higher with respect to our simulation conditions.

Appendix A: Induction Machine Parameters

3 hp, 460/230 V, 4 pole

Rotor length = 2 in

Stator slot opening = .12 in

Inner stator diameter = 4.875 in; air-gap length = 0.013 in

Number of stator slots, $ns = 36$; number of rotor slots, $nr = 44$

Stator winding configuration: single layer concentrated winding and number of coil per slot $N = 54$

Appendix B: Node Permeance Matrices

$$A_{11} = \begin{bmatrix} (Gsy_1 + Gsy_2) & -Gsy_2 & 0 & \cdots & 0 & 0 & -Gsy_1 \\ -Gsy_1 & (Gsy_2 + Gsy_3) & -Gsy_3 & \cdots & 0 & 0 & 0 \\ . & . & . & \cdots & . & . & . \\ -Gsy_1 & 0 & 0 & \cdots & 0 & -Gsy_k & (Gsy_k + Gsy_1) \end{bmatrix}$$

$$A_{22} \begin{bmatrix} \left(2G_{s\sigma} + \sum_{j=1}^{n_r} G_{1j}\right) & -G_{s\sigma} & 0 & \cdots & 0 & 0 & -G_{s\sigma} \\ -G_{s\sigma} & \left(2G_{s\sigma} + \sum_{j=1}^{n_r} G_{2j}\right) & -G_{s\sigma} & \cdots & 0 & 0 & 0 \\ . & . & & \cdots & . & . & . \\ -G_{s\sigma} & 0 & 0 & \cdots & 0 & -G_{s\sigma} & \left(2G_{s\sigma} + \sum_{j=1}^{n_r} G_{n_s j}\right) \end{bmatrix}$$

$$A_{23} = \begin{bmatrix} -G_{11} & -G_{1,2} & \cdots & -G_{1,n_r} \\ -G_{2,1} & -G_{2,2} & \cdots & -G_{2,n_r} \\ . & . & \cdots & . \\ -G_{n_s,1} & -G_{n_s,2} & \cdots & -G_{n_s,n_r} \end{bmatrix}$$

$$A_{32} = A_{23}^T$$

$$A_{33} = \begin{bmatrix} \left(G_{r\sigma,1} + G_{r\sigma,2} + \sum_{i=1}^{n_r} G_{i,1}\right) & -G_{r\sigma,2} & 0 & \cdots & 0 & 0 & -G_{r\sigma,1} \\ -G_{r\sigma,2} & \left(G_{r\sigma,2} + G_{r\sigma,3} + \sum_{i=1}^{n_s} G_{i,3}\right) & -G_{r\sigma,2} & \cdots & 0 & 0 & 0 \\ \vdots & \vdots & \vdots & \cdots & \vdots & \vdots & \vdots \\ -G_{r\sigma,1} & 0 & 0 & \cdots & 0 & -G_{r\sigma,n_r} & \left(G_{r\sigma,1} + G_{r\sigma,n_r} + \sum_{i=1}^{n_s} G_{i,n_r}\right) \end{bmatrix}$$

$$A_{44} = \begin{bmatrix} (Gry_1 + Gry_2) & -Gry_2 & 0 & \cdots & 0 & 0 & 0 \\ -Gry_2 & (Gry_2 + Gry_3) & -Gry_3 & \cdots & 0 & 0 & 0 \\ \vdots & \vdots & \vdots & \cdots & \vdots & \vdots & \vdots \\ 0 & 0 & 0 & \cdots & 0 & -Gry_{n_{r-1}} & (Gry_{n_{r-1}} + Gry_{n_r}) \end{bmatrix}$$

References

[1] ELEN 2140, Tafresh University class notes on "General Theory of Electrical Machines," Spring 2006.

[2] M. Mehravaran, "Three Dimensional Modeling and simulation of Salient Pole Synchronous Machine Using Magnetic Equivalent Circuit Method," Msc. Thesis, Tafresh University, Tafresh, Iran, Feb. 2007.

[3] H. Meshgin Kelk, "Simultaneous Three Dimensional Modeling of Squirrel Cage Induction Motor under Conventional Rotor Faults," PhD dissertation, Amirkabir University of Technology, November 2000.

[4] H. Meshgin Kelk, J. Milimonfared, and H.A. Toliyat, "A comprehensive method for the calculation of inductance coefficients of cage induction machines," *IEEE Transactions on Energy Conversion*, vol. 18, no. 2, pp. 187–193, June 2000.

[5] H. Meshgin Kelk and J. Milimonfared, "Effects of rotor air-gap eccentricity on the power factor of squirrel cage induction machines," ICEM 2002, Belgium, 2002.

[6] V. Ostovic, *Dynamics of Saturated Electric Machines*, New York: Springer-Verlag, 1989.

[7] V. Ostovic, *Computer-Aided Analysis of Electric Machines*, New York: Prentice Hall, 1994.

5

Analysis of Faulty Induction Motors Using Finite Element Method

Bashir Mahdi Ebrahimi, Ph.D.
University of Tehran

5.1 Introduction

The basis of any reliable fault diagnosis method of electrical machines is precise performance analysis of them at different conditions. Modeling of faulty machines is the first step of this procedure and has considerable effects on the accuracy of results. The features that are utilized for fault detection are extracted from processing of signals that are simulated at this stage. The modeling approaches that ignore effective characteristics of the machines cannot be used for modeling faulty machines. The two-dimensional (2-D) and three-dimensional (3-D) finite element method (FEM) as powerful simulators have been utilized to model faulty machines in different cases. In these methods, spatial distribution of the stator windings, nonuniformity of the air-gap due to stator and rotor slots, nonlinearity characteristics of the stator and rotor core materials, skin effects, skewing of the rotor bars, end effects of the stator windings and eddy currents are taken into account. Although all the aforementioned characteristics are taken into account in the 3-D FEM, some of these characteristics, such as skewing of the rotor bars and end effects of the stator windings, are not considered in the 2-D FEM. Moreover, the calculated torque using 2-D FEM is torque per length, which should be multiplied by the motor stack. In these modeling approaches, the field distribution within the machines is determined. Then, other parameters and variables of the machines such as inductances, currents, the electromotive force (EMF), developed torque, and speed of the machines are calculated. It is noticeable that symmetrical characteristics of the machines may be used to model a quarter or a half of the healthy machines instead of modeling the complete machine. However, this simplification cannot be used in the case of faulty machines.

Based on the supply to the machine, FEMs are classified into current-fed and voltage-fed approaches. In the current-fed approach, an equivalent current density is applied to the coils and then vector potential and flux density are calculated in any area of the machine. It is obvious that this method cannot be employed to compute the stator currents as the most popular signals for processing and feature extraction because on this technique, the stator currents have been supposed to be known values by their equivalent current densities. The time-stepping finite element coupled state space (TSFEM-SS) has been proposed to solve this problem. In this technique, the inductances of the machines are calculated using the current-fed FEM. Then, the resultant inductances are used in the state space equations to determine the other variables and parameters. In most cases, the voltage-fed time-stepping finite element method (TSFEM) has been utilized to calculate machine signals. In this technique, the FE area is coupled to the electrical circuits and mechanical loads. Modeling of faulty induction motors (IMs) using TSFEM has four essential parts. They are geometrical modeling, winding modeling, mechanical coupling, and fault modeling.

5.2 Geometrical Modeling of Faulty Induction Motors Using Time-Stepping Finite Element Method (TSFEM)

In order to model the geometry of an IM, all parts of the motor, which include the shaft, stator and rotor slots, stator and rotor laminations, are modeled. Then, the physical characteristics of any part of the motor are applied based on the practical materials used. For instance, in IMs, stator slots are filled by copper, which has evident permeability and conductivity. The rotor slots, which are filled by aluminum with known permeability and conductivity, are short circuited. The B-H curve of the materials used in the stator and rotor cores is taken into account. Figure 5.1 depicts the 2-D stator and rotor laminations of the IM. The 3-D configuration of the same motor has been demonstrated in Figure 5.2.

According to Figure 5.1 and Figure 5.2, there are some differences between 2-D and 3-D TSFEMs. It is seen that end effects of the stator windings have been taken into account in 3-D modeling. This characteristic

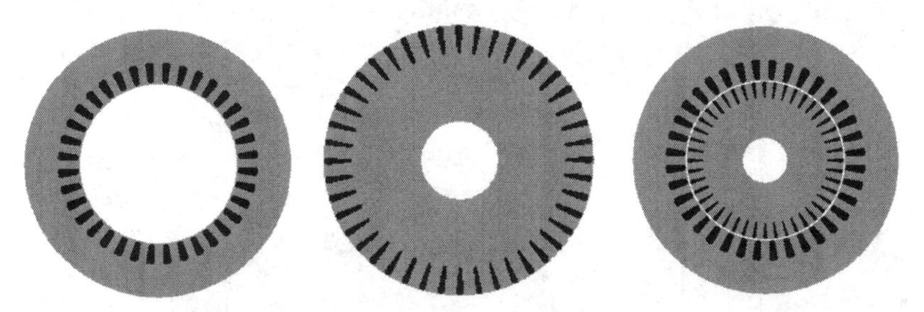

FIGURE 5.1
Cross-section of (left) stator, (middle) rotor, and (right) whole motor.

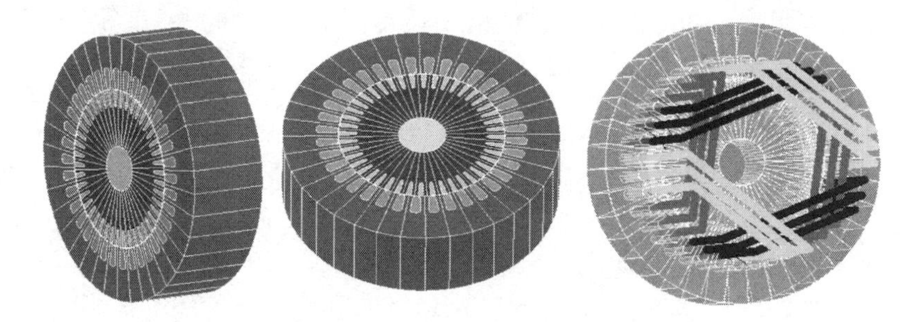

FIGURE 5.2
The 3-D configuration of the motor illustrated in Figure 5.1.

can be modeled in the 2-D FEM using constant inductance in the electrical circuits, which are coupled to the finite element (FE) area. Moreover, the skin effects are taken into account in the 3-D modeling. Nonetheless, considering skin effects in the 2-D modeling depends on the user experience in applying mesh. The skewing of the rotor bars is taken into account using 3-D modeling, whereas it is ignored in the 2-D modeling. Therefore, the simulated torque has more ripples than that in the 2-D modeling. The influence of the motor stack is reckoned in the 3-D modeling, whereas a cross-section of the motor is simulated in the 2-D modeling (see Figure 5.1).

5.3 Coupling of Electrical Circuits and Finite Element Area

This stage of modeling procedure has considerable impact on the simulation results accuracy. The sinusoidal or nonsinusoidal supply types are determined here. In this stage, the motor is fed by the three-phase sinusoidal supply, unbalanced sinusoidal supply, or inverters. Figure 5.3 illustrates the

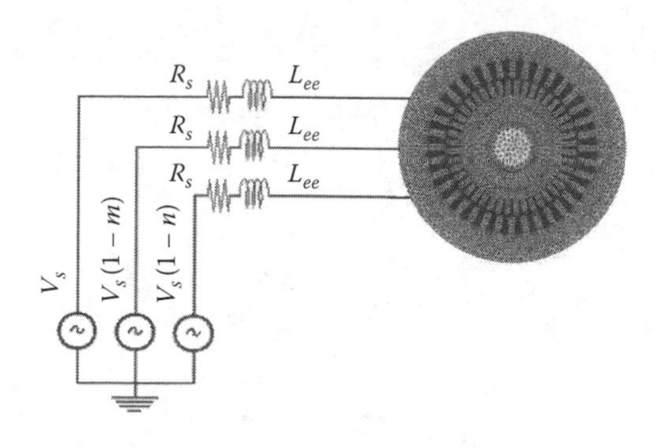

FIGURE 5.3
Coupling electrical circuits to the finite element area.

coupling between electrical circuits and FE area in different supply conditions. According to Figure 5.3, the end effects of the stator windings have been modeled using external inductances, which are calculated analytically and added to the electrical circuits. The transient equations of the external circuit that exhibits the electric supplies and circuit elements are combined to the field equations in FEM. Also, the motion equations due to mechanical coupling are combined to the mentioned previous electromagnetic equations. Solution of these equations yields the magnetic flux density distribution, the stator phase current, the EMF, the developed torque, and speed of the motor. Two-dimensional magnetic field propagation is given as follows:

$$\frac{\partial}{\partial x}\frac{1}{\mu}\left(\frac{\partial A}{\partial x}\right)+\frac{\partial}{\partial y}\frac{1}{\mu}\left(\frac{\partial A}{\partial y}\right)=J_0+J_e+J_v \tag{5.1}$$

where A is the z-component of the magnetic vector potential, and μ is the magnetic permeability. J_0 is the current density related to the applied voltage, J_e is the current density related to the time variations of the magnetic flux, and J_v is the current density related to the motional voltage. Therefore, Equation (5.1) is rewritten as follows:

$$\frac{\partial}{\partial x}\frac{1}{\mu}\left(\frac{\partial A}{\partial x}\right)+\frac{\partial}{\partial y}\frac{1}{\mu}\left(\frac{\partial A}{\partial y}\right)=-\sigma\frac{V_s}{\ell}+\sigma\frac{\partial A}{\partial t}+\sigma v\times\nabla\times A \tag{5.2}$$

where σ is the electrical conductivity, ℓ is the motor stack along z-axis, V_s is the applied voltage, and v is the speed of the conductor against magnetic flux density. By applying a reference frame that is assumed fixed in respect to the proposed element, v is equal to zero, and the propagation equation is simplified as follows:

$$\frac{\partial}{\partial x}\frac{1}{\mu}\left(\frac{\partial A}{\partial x}\right)+\frac{\partial}{\partial y}\frac{1}{\mu}\left(\frac{\partial A}{\partial y}\right)=-\sigma\frac{V_s}{\ell}+\sigma\frac{\partial A}{\partial t} \tag{5.3}$$

The circuit equation of the magnetic coil is given as follows:

$$V_s(t)=R_s i_s(t)+L_{ee}\frac{di_s(t)}{dt}+emf(t) \tag{5.4}$$

where R_s is the stator resistance, i_s is the stator phase current, L_{ee} is the external inductance added to the electrical circuits due to end effects of the stator windings, and *emf* is the applied voltage to the FE area. By coupling Equation (5.3) and Equation (5.4), the TSFEM is used to obtain the magnetic vector potential, stator currents, and the EMF. The nonlinear equation that can relate the FE equations expressing the electromagnetic fields of the machine with the circuit equations is as follow:

$$[C][A \quad emf \quad i_s]^T + [D]\left[\frac{\partial A}{\partial t} \quad \frac{\partial emf}{\partial t} \quad \frac{\partial i_s}{\partial t}\right]^T = [P] \tag{5.5}$$

where $[C]$ and $[D]$ are the coefficients matrices, $[P]$ is the vector related to the input voltage, and the solution of Equation (5.5) gives $[A]$ and $[i_s]$ as essential signals for analyzing and processing.

5.4 Modeling Internal Faults Using Finite Element Method

Internal faults in IMs are categorized to broken bars as an electrical fault; bearing fault, and static eccentricity (SE), dynamic eccentricity (DE), and mixed eccentricity (ME) as mechanical faults.

5.4.1 Modeling Broken Bar Fault

In the study by Elkasbagy et al. [1], the bar current is taken to be equal to zero for modeling broken bars. It is noted that zero current in a particular bar increases the currents of the adjacent bars considerably. This implies a considerable asymmetry in the rotor circuit and consequently asymmetry in the field produced by the rotor currents. Even a broken bar can have a nonzero current, depending on the type of construction and the way aluminum is used in manufacturing. In fact, current paths exist between the bars of the squirrel-cage rotor. For instance, currents can enter the bar where it is connected to the end-ring and return through the rotor core. Additional current paths are created because of injection of the high-pressure molten aluminum due to injection of aluminum into the rotor core in the manufacturing procedure. This molten aluminum penetrates between the sheets, which can generate a conducting path between two adjacent bars. Therefore, in the modeling of broken bars, currents are considered nonzero, but at the same time the resistance of the broken bar is taken to be high. For example, Faiz and Ebrahimi considered the resistance of the healthy IM as 39.42 μΩ and the broken bar as 2500 μΩ [2]. Figure 5.4 reveals the flux distribution within the healthy and faulty IM under one broken bar. It is seen that bar breakage distorts flux distribution. Furthermore, a local saturation is observed due to current flow of the broken bar in adjacent bars. Another important point that should be taken into account in the diagnosis of the rotor broken bars is locating broken bars. As shown in Figure 5.5, eight cases can be imagined when

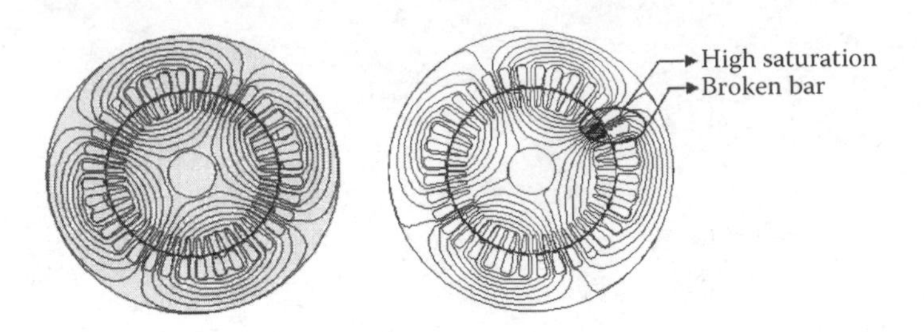

FIGURE 5.4
Magnetic flux distribution for induction motor: (left) healthy and (right) one broken bar.

there are four broken bars in the rotor cage, and such case has a considerable effect upon the amplitude of the harmonic components due to the fault.

Figure 5.6 depicts the time variation of the stator currents in the healthy and faulty IM with four broken bars. It is seen that bars breakage unbalances the stator current profile. Furthermore, envelope ripples of the stator current profile in the faulty case is more than that healthy case.

Figure 5.7 demonstrates torque profiles of the healthy and faulty IM with four broken bars. It is observed that torque ripples of the faulty case are more considerable than those of the healthy case. This is due to distortion of the

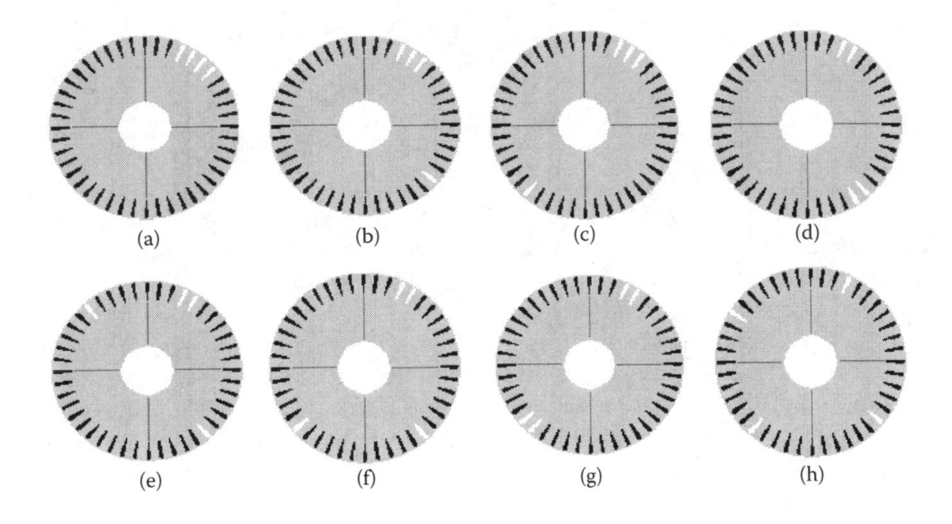

FIGURE 5.5
Different cases for distribution of broken bars under poles: (a) four broken bars on one pole, (b) three broken bars on one pole and one broken bar on adjacent pole, (c) three broken bars on one pole and one broken bar on opposite pole, (d) two broken bars on one pole and two broken bars on adjacent pole, (e) two broken bars on one pole and two broken bars on two adjacent poles, (f) two broken bars on one pole and one broken bar on adjacent and opposite poles, (g) two broken bars on one pole and two broken bars on opposite pole, and (h) one broken bar on each pole.

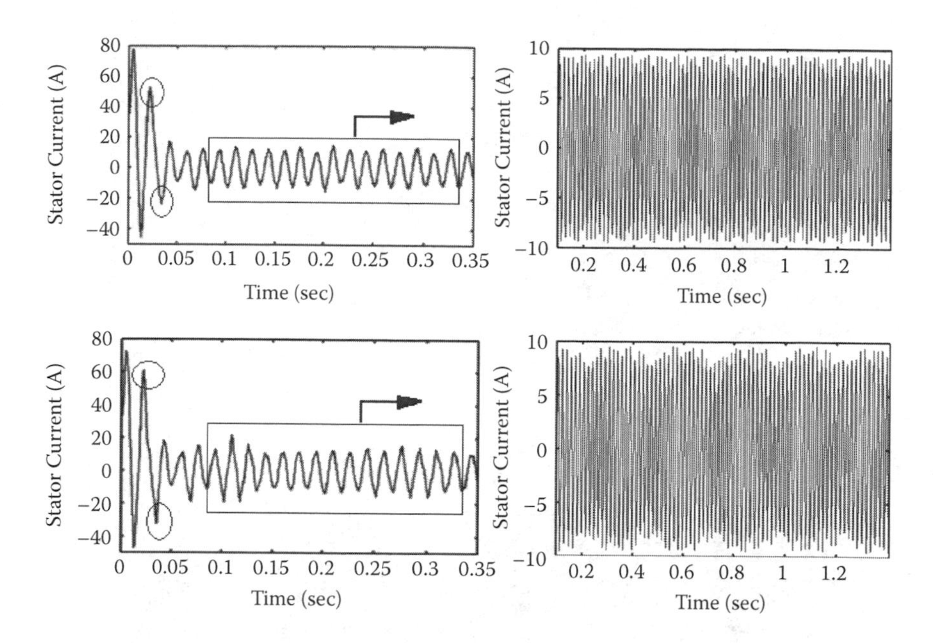

FIGURE 5.6
Time variations of current of a full-load induction motor (top) healthy and (bottom) faulty with four broken bars.

flux density distribution in the faulty case, which increases amplitude of the harmonic components of the flux density profile.

5.4.2 Modeling Eccentricity Fault

Eccentricity fault is due to bearings fatigue, manufacturing and assembling processes, and other mechanical reasons. In this fault, conformity of the stator axis, rotor axis, and rotor rotating axis are disturbed.

5.4.2.1 Static Eccentricity

In the case of static eccentricity, the rotational axis of the rotor is identical to its symmetrical axis but has been displayed with respect to the stator symmetrical axis. Although the air-gap distribution around the rotor is not uniform, it is time independent. The static eccentricity degree (δ_{se}) is defined as follows:

$$\delta_{se} = \frac{|O_s O_w|}{g_0} \tag{5.6}$$

where O_S is the stator symmetry center, O_w is the rotor rotation center, and g_0 is the uniform air-gap length. Figure 5.8 illustrates the position of stator and

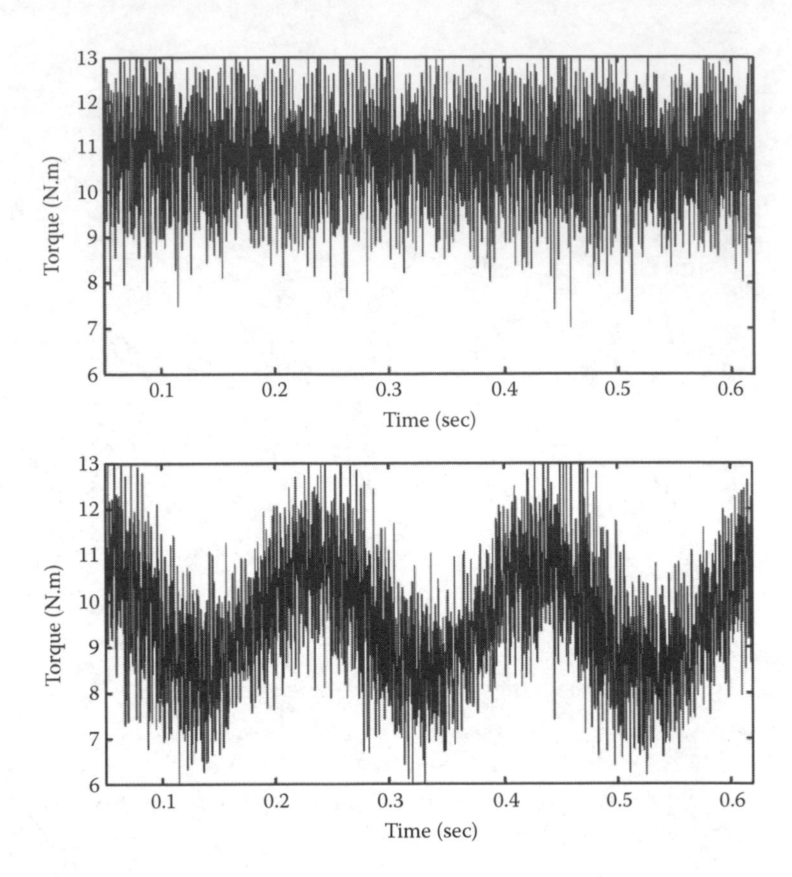

FIGURE 5.7
Steady-state time variations of current of a full-load induction motor (top) healthy and (bottom) faulty with four broken bars.

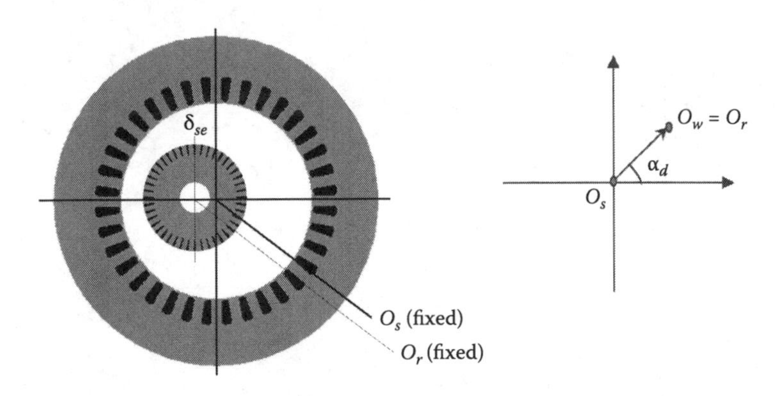

FIGURE 5.8
Geometric configurations of the modeled motor under static eccentricity.

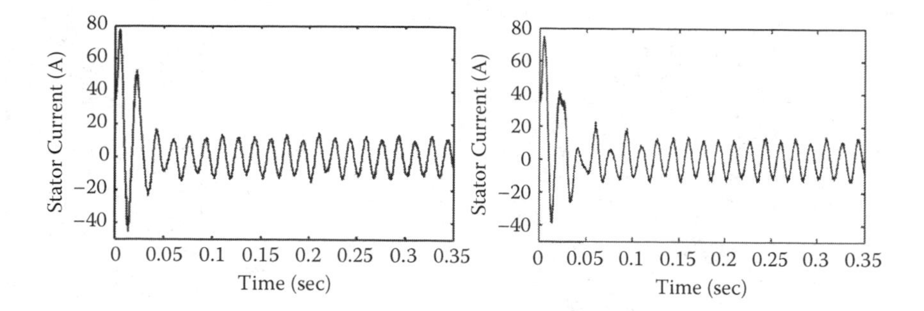

FIGURE 5.9
Time variations of stator current of a full-load induction motor: (left) healthy, (right) under 10% dynamic eccentricity.

rotor cross-sections in the static eccentricity, where α_s is the initial angle of static eccentricity and vector O_sO_w is the static transfer vector. This vector is fixed for all angular positions of the rotor. The reasons for increasing the eccentricity are bad position of the stator core due to the mounting of the motor. and nonorientation of the stator and rotor centers during the primary maintenance.

The time variation of the faulty IM under 10% static eccentricity has been exhibited in Figure 5.9. The distortion of stator currents' profiles due to static eccentricity fault is clearly seen. Figure 5.10 reveals the time variation of the torque profile in the faulty IM with 10% static eccentricity. Comparison between the health and faulty cases shows that torque ripples rise considerably in the IM under static eccentricity.

5.4.2.2 Dynamic Eccentricity

For dynamic eccentricity, the minimum air-gap length depends on the rotor angular position, and it rotates around the rotor. This may be due to

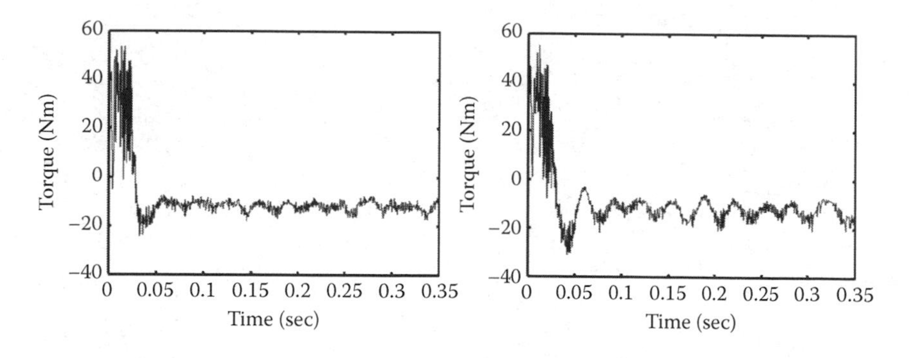

FIGURE 5.10
Time variations of developed torque of a full-load induction motor: (left) healthy, (right) under 40% dynamic eccentricity.

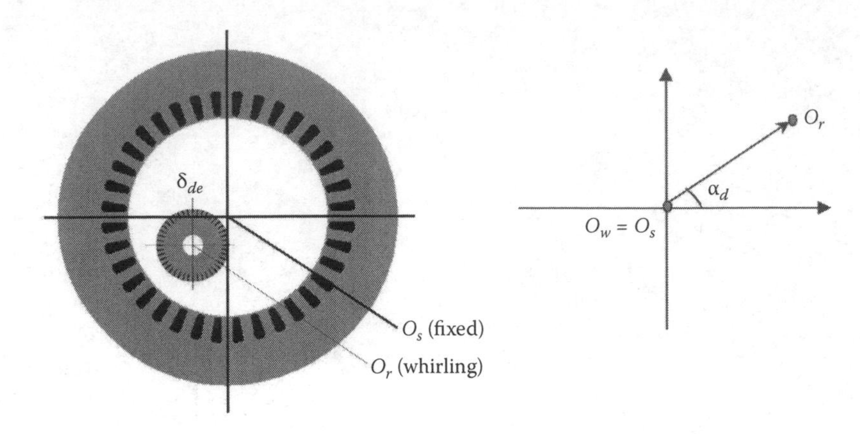

FIGURE 5.11
Geometric configurations of the modeled motor under dynamic eccentricity.

misalignment or curvature of the rotor axis. Albeit, the static eccentricity generates asymmetrical magnetic pull, which results in the dynamic eccentricity. In this eccentricity, the symmetry axis of the stator and rotation axis of the rotor is identical, but the rotor symmetry axis has been displaced. In such a case, the air gap around the rotor is nonuniform and time varying. The dynamic eccentricity degree (δ_{de}) is defined as follows:

$$\delta_{de} = \frac{|O_w O_r|}{g_0} \tag{5.7}$$

where O_r is the rotor symmetrical axis and vector $O_w O_r$ is the dynamic transfer vector. This vector is fixed for all angular positions of the rotor, but its angle varies. Figure 5.11 shows the dynamic eccentricity where α_d is the initial angle of the dynamic eccentricity.

5.4.2.3 Mixed Eccentricity

In mixed eccentricity, the symmetry axis of the rotor and stator, and the rotation axis of the rotor are displaced. This is the result of application of the resultant vector of static and dynamic transfer vectors. If there are both static and dynamic eccentricities, the eccentricity is called a mixed eccentricity. The mixed eccentricity degree (δ_{me}) is defined as follows:

$$\delta_{me} = \frac{|O_s O_r|}{g_0} = \frac{|O_s O_w + O_w O_r|}{g_0} \tag{5.8}$$

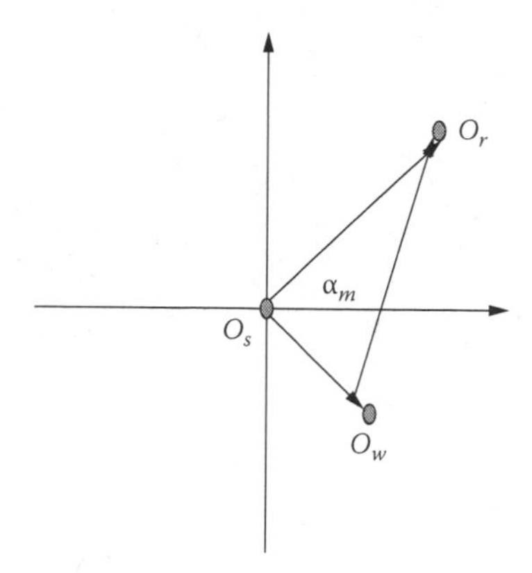

FIGURE 5.12
Position of stator and rotor mixed eccentricity in the stator reference frame.

where vector O_sO_r is the mixed transfer vector. Figure 5.12 shows that the amplitude and angle of this vector depend on the mechanical angle of the rotor, where α_m is the transfer angle of the mixed eccentricity.

5.5 Impact of Magnetic Saturation on Accurate Fault Detection in Induction Motors

One of the most effective parameters on accurate fault detection in IMs is considering nonlinear characteristics of the core materials. This characteristic is ignored in most analytical modeling methods. In this part, influence of nonlinear characteristics of the core materials on results accuracy is investigated in detail. Hence, performance of an IM that has been modeled using TSFEM is investigated in cases that permeability of the core materials has been supposed constant and also the B-H curve of the core materials has been considered. Figure 5.13 and Figure 5.14 illustrate time variation of the stator current in the healthy and faulty IM with 40% static eccentricity. In these simulations, at first permeability of the core materials has been considered constant, and in another simulation, the actual B-H curve of the core materials has been taken into account. According to Figure 5.13 and Figure 5.14, the variation rate of the stator current in the transient mode where permeability is constant is much more than the case in which nonlinear characteristics of the core materials have been considered. This variation rate increases when the fault occurs. Furthermore, the necessary time to reach the steady state in the model with constant permeability is very large in comparison with the modeled IM considering magnetization curve.

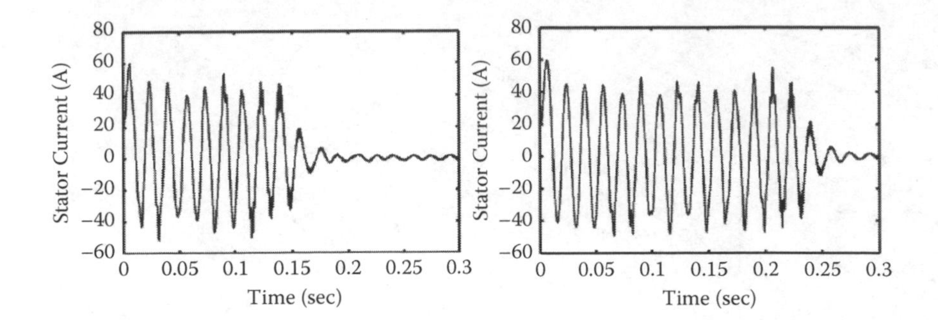

FIGURE 5.13
Time variations of stator current of a no-load induction motor with constant permeability: (left) healthy, (right) 40% static eccentricity.

Figure 5.15 and Figure 5.16 exhibit the time variation rate of the torque profile of the healthy and faulty motor in the two aforementioned modeling methods. It is seen that torque ripples in the modeled IM with constant permeability is much more than the case in which non-linear characteristics of the core materials have been taken into account. These ripples rise noticeably when fault occurs.

Figure 5.17 reveals the time variation rate of the rotor speed of the healthy and faulty motor in a modeled motor with and without magnetization curve. It is seen that speed ripples in the transient mode of the modeled IM with constant permeability is much more than the case in which nonlinear characteristics of the core materials have been taken into account. These ripples rise noticeably when fault occurs. Furthermore, the settling time of the motor speed of the modeled IM with constant permeability is much more than the other case. In order to justify these results, air-gap flux density of the healthy and faulty IM in the transient and steady-state modes is studied.

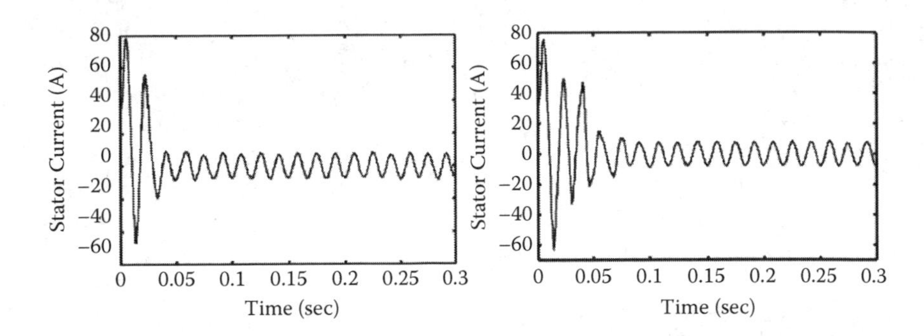

FIGURE 5.14
Time variations of stator current of a no-load induction motor considering magnetizing curve: (left) healthy, (right) 40% static eccentricity.

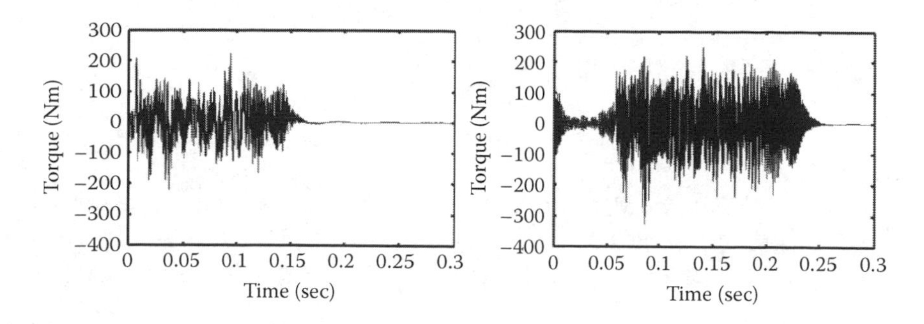

FIGURE 5.15
Time variations of torque profile of a no-load induction motor with constant permeability: (left) healthy, (right) 40% static eccentricity.

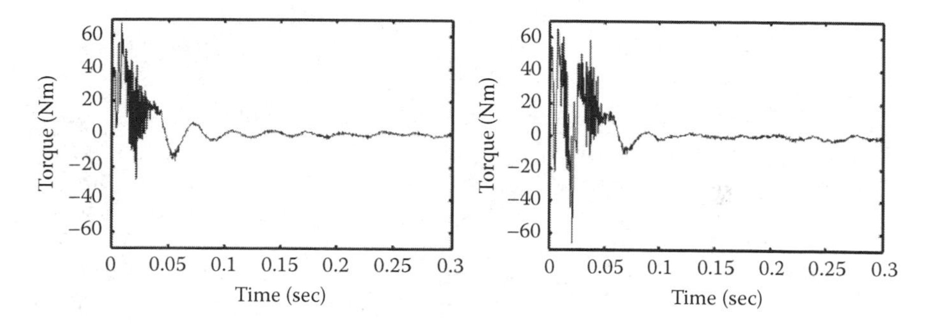

FIGURE. 5.16
Time variations of torque profile of a no-load induction motor considering magnetizing curve: (left) healthy, (right) 40% static eccentricity.

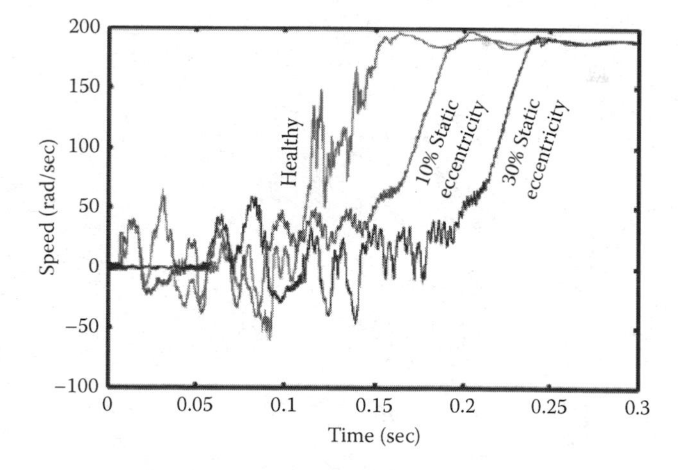

FIGURE 5.17
Time variations of rotor speed of a no-load induction motor with constant permeability in healthy and different faulty cases.

5.5.1 Analysis of Air-Gap Magnetic Flux Density in Healthy and Faulty Induction Motor

The magnetic field waveform contains full information of the stator condition and mechanical parts of the motor. Therefore, it is possible to predict and diagnose all faults by continuous monitoring of the air-gap magnetic field. In order to measure the air gap magnetic field practically, a search coil is used. This coil is placed inside the stator slots. By integration of the terminal voltage of this coil, the air-gap magnetic field is determined.

5.5.1.1 Linear Magnetization Characteristic

Figure 5.18 and Figure 5.19 present the air-gap magnetic flux density for a healthy and faulty motor at the startup and steady-state modes when magnetization characteristic is assumed to be linear. Figure 5.18a shows the corresponding characteristic at the start-up of the healthy induction motor. Figure 5.18b shows the air-gap magnetic flux density under 30% static eccentricity at the startup. Comparison of Figure 5.18a and Figure 5.18b indicates that the static eccentricity unrealistically increases the magnetic flux density of the air-gap. The reason is analysis of the motor performance is based on the fixed permeability. Also, comparison of Figure 5.18a and Figure 5.18b indicates that the static eccentricity leads to asymmetry of the magnetic flux density distribution, because development of the fault generates new harmonic components in the air-gap field. Figure 5.19 exhibits the air-gap magnetic flux density of the healthy and faulty induction motor in the steady-state mode. It is necessary to mention that analysis of the induction motor using the linear magnetization characteristic shows a much larger air-gap magnetic flux density. This is clearer at the start-up mode. The reason is that during start-up of the induction motor, slip varies largely, and the machine currents are very large. So, the magnetic flux density rises linearly according to the constant

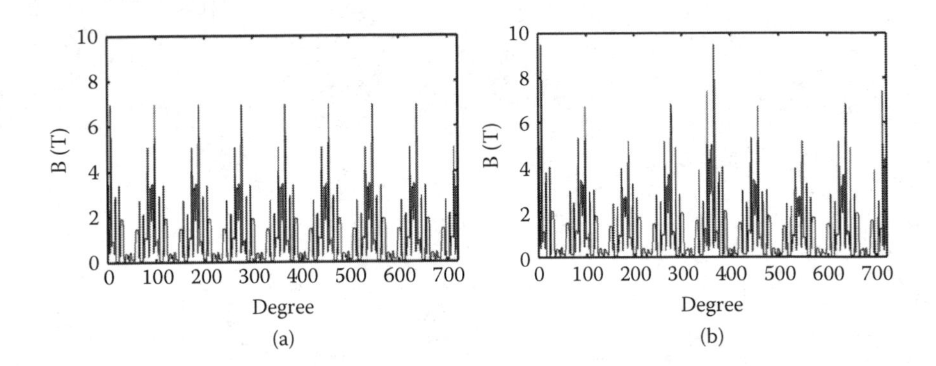

FIGURE 5.18
Flux density distribution in transient mode of induction motor air-gap with linear magnetization characteristic: (a) healthy, (b) 30% static eccentricity.

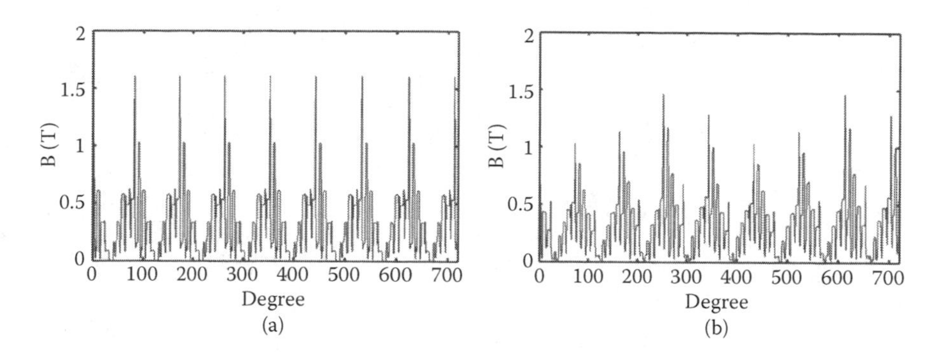

FIGURE 5.19
Flux density distribution in steady-state mode of induction motor air-gap with linear magnetization characteristic: (a) healthy, (b) 30% static eccentricity.

permeability; since the saturation has been ignored, the rate of the flux density variations during startup is very large, and amplitude of the air gap magnetic flux density is considerable (Figure 5.18). If SE occurs during the startup period or a faulty motor is started, amplitude of the magnetic flux density increases for the faulty induction motor, so amplitude of the magnetic flux density of the faulty induction motor increases during the startup compared with that of the healthy motor (Figure 5.18b). Referring to Figure 5.19, amplitudes of the air-gap magnetic flux densities of a healthy and faulty motor, under steady-state mode, decreases compared with that of the start-up mode. However, a linear magnetization characteristic has been used in this analysis, and amplitude of the air-gap magnetic flux density is larger than the actual case.

5.5.1.2 Nonlinear Magnetization Characteristic

Figure 5.20 reveals the air-gap magnetic flux density distribution of a healthy and faulty induction motor at the startup using the no-linear magnetization

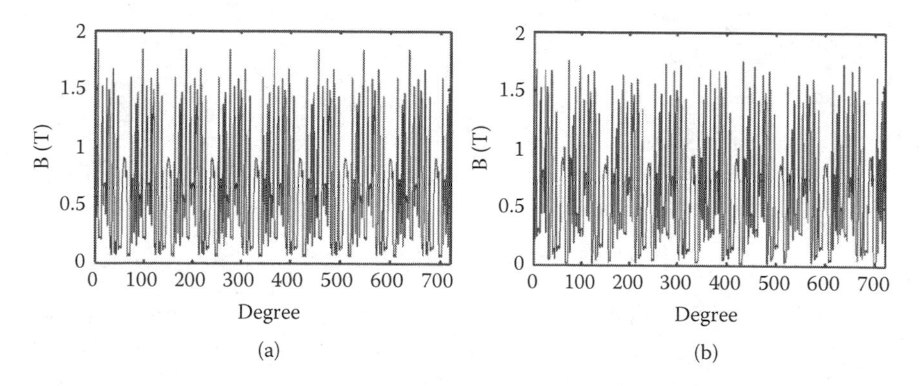

FIGURE 5.20
Flux density distribution in transient mode of induction motor air-gap considering magnetizing curve: (a) healthy, (b) 30% static eccentricity.

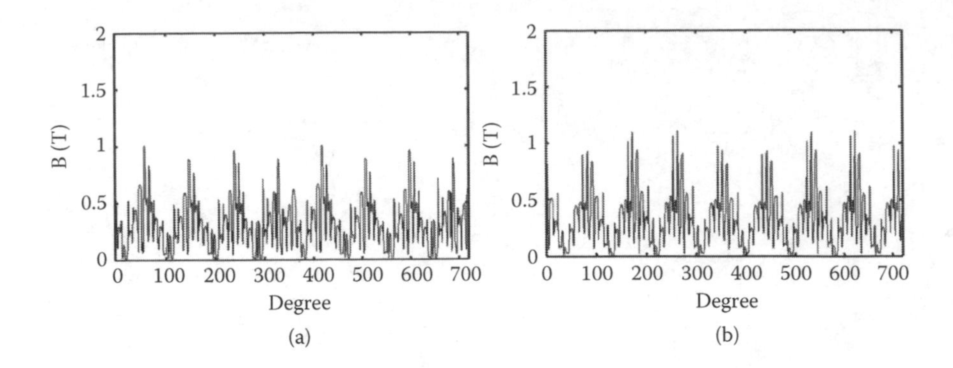

FIGURE 5.21
Flux density distribution in steady-state mode of induction motor air gap considering magnetizing curve: (a) healthy, (b) 30% static eccentricity.

characteristic. Comparison of Figure 5.20a and Figure 5.20b indicates the asymmetry of the magnetic flux density distribution of the faulty motor with 30% SE. An important point in Figure 5.20 is the considerable decrease of the air-gap magnetic flux density during start-up while the motor is analyzed using the nonlinear magnetization characteristic. In spite of this, the reason for large amplitude of the magnetic flux density observed in Figure 5.20 is that in the transient analysis of the healthy and faulty induction motor, the operating point of the motor places on the linear section of the magnetization curve. After approaching the steady-state, current, speed, and slip of the motor find their nominal values. Therefore, the magnetic flux density of the healthy motor is symmetrical according to Figure 5.21, which shows the asymmetrical air-gap magnetic flux density distribution for the induction motor under 30% SE.

References

[1] N.M. Elkasbagy, A.R. Eastham, and G.E. Dawson, "Detection of broken bars in the cage rotor on an induction machine," *IEEE Transactions on Industry Applications,* vol. 22, no. 6, pp. 165–171, 1992.

[2] J. Faiz and B.M. Ebrahimi, "A new pattern for detection of broken rotor bars in induction motors during start-up," *IEEE Transactions on Magnetics,* vol. 44, no. 12, pp. 4673–4683, December 2008.

[3] J. Faiz, B.M. Ebrahimi, B. Akin, and H.A. Toliyat, "Dynamic analysis of mixed eccentricity signatures at various operating points and scrutiny of related indices for induction motors," *IET Electric Power Applications,* vol. 4, no. 1, pp. 1–16, February 2010.

[4] J. Faiz, B.M. Ebrahimi, and H.A. Toliyat, "Effect of magnetic saturation on static and mixed eccentricity fault diagnosis in induction motors," *IEEE Transactions on Magnetics*, vol. 45, no. 8, pp. 3137–3144, August 2009.
[5] J. Faiz and B.M. Ebrahimi, "Locating rotor broken bars in induction motors using finite element method," *Journal of Energy Conversion and Management*, vol. 50, pp. 125–131, 2009.
[6] J. Faiz, B.M. Ebrahimi, B. Akin, and B. Asaie, "Criterion function for broken-bar fault diagnosis in induction motor under load variation using wavelet transform," *Journal of Electromagnetics*, vol. 29, pp. 220–234, May 2009.
[7] J. Faiz, B.M. Ebrahimi, B. Akin, and H.A. Toliyat, "Finite-element transient analysis of induction motors under mixed eccentricity fault," *IEEE Transactions on Magnetics*, vol. 44, no. 1, pp. 66–74, January 2008.
[8] J. Faiz and B.M. Ebrahimi, "Determination of number of broken rotor bars and static eccentricity degree in induction motor under mixed fault," *Journal of Electromagnetics*, vol. 28, pp. 433–449, August 2008.
[9] J. Faiz and B.M. Ebrahimi, "Influence of magnetic saturation upon performance of induction motor using time stepping finite element method," *Electric Power Components and Systems*, vol. 27, 505–524, 2007.

6

Fault Diagnosis of Electric Machines Using Techniques Based on Frequency Domain

Subhasis Nandi, Ph.D.
University of Victoria

6.1 Introduction

Generally speaking, most of the fault detection techniques used in real-time fault detection in power systems are time-domain based. The over current, over voltage, earth fault, impedance relays, and so forth are mostly time-domain based. However, as far as detecting faults for electric machines are concerned, frequency-domain-based techniques, especially ones based on fast Fourier transforms (FFT) are very popular. Except for stator-related faults, most other faults can be reliably diagnosed using a spectrum analyzer provided the machines are operating under steady-state conditions for at least a reasonable period of time. For applications in which machines are made to operate under very frequently fluctuating load and speed conditions, traditional FFT has to be replaced with short time Fourier transforms (STFT), spectrograms, and other time-frequency analysis using wavelets and Wigner–Ville transforms. Usually the machine current, flux, mechanical vibration, torque, and speed signals are analyzed in frequency domain. High-resolution spectral techniques such as multiple signal classification (MUSIC), ROOTMUSIC, and higher-order spectral methods such as bispectrum and trispectrum have also been proposed by a few researchers. However, most of the popular frequency-domain-based techniques are based on fast Fourier transform of the line current generally known as motor current signature analysis (MCSA). Sometimes, when the frequencies at which the detections are to be made are known, swept sine measurements or the digital frequency locked loop technique (DFLL) are also used. This avoids lengthy computations while achieving good resolution.

Traditionally, in many countries, power engineers are not exposed even to the basic signal processing course, which is only taught to students in electronics and communication. Hence it will not be out of place to discuss a few basics of signal processing first, before going into actual fault diagnosis using signal processing.

6.2 Some Definitions and Examples Related to Signal Processing

6.2.1 Continuous versus Discrete or Digital or Sampled Signal

A *continuous signal*, $x(t)$, is one that is defined at any given point of time. Examples of such a signal are the line currents and line voltages of a motor that we can observe on an analog oscilloscope. The same signal data when acquired through a data acquisition system or seen through a digitizing oscilloscope becomes a *discrete signal*, $x(n)$, which is nothing but the sampled version of the continuous signal at a regular time interval, T_{sp} [1]. The frequency at which the sampling device works is f_{sp}. In general,

$$T_{sp} = \frac{1}{f_{sp}} \tag{6.1}$$

$$x(n) = x(t)\big|_{t=nT_{sp}} \tag{6.2}$$

Often one has to prefilter a signal to avoid *aliasing* (literally meaning "same name for one thing") arising out of the sampling process. Unless proper care is taken in choosing the sampling frequency or the prefilter, one frequency component may be wrongly interpreted as another while trying to determine the frequency components present in a discrete signal. It must be noted, though, that sometimes aliasing can be used beneficially, too, such as in a stroboscope, a device used to measure speed, or for strengthening weak signals used for fault detection.

Example 6.1

Suppose we have a voltage signal given by $x(t)=100\sin(2\pi 60t)+10\sin(2\pi 300t)$. Determine how the sampled version of this signal would look like when sampled by (a) a 200 Hz signal and (b) a 1000 Hz signal.

a. We have, using Equation (6.2)

$$\begin{aligned} x(n) &= 100\sin(2\pi 0.3n) + 10\sin(2\pi 1.5n) \\ &= 100\sin(2\pi 0.3n) + 10\sin(2\pi n + 2\pi 0.5n) \\ &= 100\sin(2\pi 0.3n) + 10\sin(2\pi 0.5n) \end{aligned}$$

Now had the signal been $x(t)=100\sin(2\pi 60t)+10\sin(2\pi 100t)$, the result would have been the same, meaning that with 200 Hz sampling frequency the 300 Hz signal can be misconstrued as a 100 Hz signal.

b. However in this case $x(n)=100\sin(2\pi 0.06n)+10\sin(2\pi 0.3n)$ and the 300 Hz signal can be easily distinguished from a 100 Hz signal.

In order for proper signal reconstruction or interpretation after sampling, a continuous time signal $x(t)$ has to be sampled at a rate greater than twice the maximum frequency contained in that signal. This result is actually one of the fundamental theorems in signal processing and is known as the *Shannon sampling theorem.*

6.2.2 Continuous, Discrete Fourier Transforms, and Nonparametric Power Spectrum Estimation

A *continuous Fourier transform* is given by the following two formulas [1–3]:

$$X(j\omega) = \int_{-\infty}^{\infty} x(t)e^{-j\omega t}\, dt \tag{6.3}$$

$$x(t) = \frac{1}{2\pi}\int_{-\infty}^{\infty} X(j\omega)e^{j\omega t}\, d\omega \tag{6.4}$$

Equation (6.3) is known as the *analysis* or *forward* equation because it extracts the frequency information from the time-domain signal. Equation (6.4) is known as the *synthesis* or *inverse* equation because it creates the original time-domain signal back from the spectral information.

A *discrete Fourier transform* (DFT), on the other hand, is given by [1–3]:

$$X(k) = \sum_{n=0}^{N-1} x(n)e^{-\frac{j2\pi nk}{N}}, \quad k = 0,1,\ldots\ldots, N-1 \tag{6.5}$$

$$x(n) = \frac{1}{N}\sum_{k=0}^{N-1} X(k)e^{\frac{j2\pi nk}{N}}, \quad n = 0,1,\ldots\ldots, N-1 \tag{6.6}$$

with Equation (6.5) analogous to Equation (6.3) and Equation (6.4) analogous to Equation (6.6). It is also interesting to note that while both Equation (6.3) and Equation (6.4) are in continuous domain, both Equation (6.5) and Equation (6.6) are in discrete domain. It is also possible to write the analysis equation in the discrete form *but* the synthesis equation in the continuous form. In this case the equation set is known as *discrete-time Fourier transform,* meaning the discretization is done in time-domain only. The other alternative—that is, the equation set with the analysis equation in continuous form and the synthesis equation in discrete form—is the very well known form of the *Fourier series* that essentially expresses a periodic but continuous time in terms of discrete frequency components. Although it is not difficult to show the relationship between the four aforementioned kinds of transforms, the DFT is the most important for the fault diagnosis purpose. At this point in time it also becomes necessary to revisit the definition of FFT. A FFT is nothing but the collection of algorithms used for efficient computation of the DFT.

Example 6.2: Compute the Continuous Fourier Transform of the Voltage Signal Given in Example 6.1

Let us begin by computing the time-domain signal corresponding to the impulse signal in frequency domain given by

$$X(j\omega) = \delta(\omega - \omega_0) \tag{6.7}$$

Using Equation (6.4)

$$x(t) = \frac{1}{2\pi}\int_{-\infty}^{\infty} \delta(\omega - \omega_0)e^{j\omega t}d\omega = \frac{e^{j\omega_0 t}}{2\pi}\int_{-\infty}^{\infty} \delta(\omega - \omega_0)d\omega = \frac{e^{j\omega_0 t}}{2\pi} \tag{6.8}$$

The signal in this case, using Euler's identity can be written as

$$y(t) = 100\sin(2\pi 60t) + 10\sin(2\pi 300t) = \frac{100}{2j}(e^{j2\pi 60t} - e^{-j2\pi 60t}) + \frac{10}{2j}(e^{j2\pi 300t} - e^{-j2\pi 300t}) \tag{6.9}$$

Hence using Equation (6.7) and Equation (6.8)

$$Y(j\omega) = \frac{100\pi}{j}(\delta(\omega - 2\pi 60t) - \delta(\omega + 2\pi 60t)) + \frac{10\pi}{j}(\delta(\omega - 2\pi 300t) - \delta(\omega + 2\pi 300t)) \tag{6.10}$$

Note that in Equation (6.10) for each frequency component there are two delta functions. Though their phases are opposite, their magnitudes are same. Normally one is interested in the magnitude and therefore it is sufficient to have only one of the delta functions. Conventionally, only those lying on the right side of $\omega = 0$ (that is, $\delta(\omega - 2\pi 60t)$ and $\delta(\omega - 2\pi 300t)$ in this example) are chosen.

Since one has to work with a finite data set in practice, DFT is the transform to be used. If one could acquire a large set of steady-state data with minimal temporal variations, DFT signals would approach the true single line nature of the spectra as given by Equation (6.9) in a limiting sense. Unfortunately, many times, due to constraints such as speed of computation and memory, one has to use a limited data set. Significant improvement, however, in the quality of the signal can be obtained by judicious choice of the *window function*. The limited data set is analogous to viewing something through a small window. Now if the glass pane on the window is not clear enough, details of whatever is viewed may not be distinct. The simplest window, as one could intuitively guess, is the so- called *rectangular window,* embedded due to the very fact that the data set is limited in nature. The rectangular window has continuous spectra and as a result the power of the original signal data, instead of being concentrated at the points of interest, leaks out over the entire frequency range. This is called *spectral leakage.* Specialized window functions such as the Hanning window and Bartlett window are used to reduce spectral leakage. However, windows result in a loss of resolution. The only way to improve resolution is to increase N in Equation (6.5) and Equation (6.6). This can be achieved only by increasing the window length, meaning increasing the length of the data set. Increasing sampling frequency will not improve the resolution of the spectrum.

In practice, any data is bound to have some noise associated with it. As long as the noise is white (zero mean, unit variance), it can be easily minimized by averaging several *power spectral density* (PSD) spectra as given by Equation (6.10)

$$X(\frac{k}{N}) = \frac{1}{N}\left|\sum_{n=0}^{N} x(n)e^{-\frac{j2\pi nk}{N}}\right|^2, k = 0,1,......,N-1 \tag{6.11}$$

computed over small data segments. This is essentially computing the square of the magnitude of the FFT of the segments and then averaging them. In general this method is known as *nonparametric spectrum estimation* [1–3]. These segments can be either overlapping or nonoverlapping. For a given data set, the noise reduction is attained at the cost of frequency resolution and vice versa. Depending upon the type of averaging techniques or window used, the nonparametric power spectrum estimation may be known as, for example, the *periodogram, Bartlett, Welch,* or *Blackman-Tukey.*

Example 6.3

a. Suppose a 1-second data set of the signal given by $x(t)$=100sin($2\pi 60t$)+2sin($2\pi 63t$)+2sin($2\pi 57t$)+10sin($2\pi 180t$)+*white noise.* The signal has been obtained with a sampling frequency of 3600 Hz. The *white noise* used is random numbers that vary between 0 and 1. From the machine diagnostic viewpoint such a signal would approximately represent the low frequency spectrum of an induction motor with broken rotor bars. Plot the FFT using all 3600 points with (i) a Rectangular and (ii) a Hanning window.
b. Suppose a 10-second data set is used. Repeat (a).
c. Plot the PSD spectrum of the 10-second data set, with segment size of 12,000 data points and 10,000 overlapping data points between two segments. Use Hanning window.

Comparing the first two plots in Figure 6.1, the reduction of resolution with Hanning window is quite clear. However the spectral leakage is

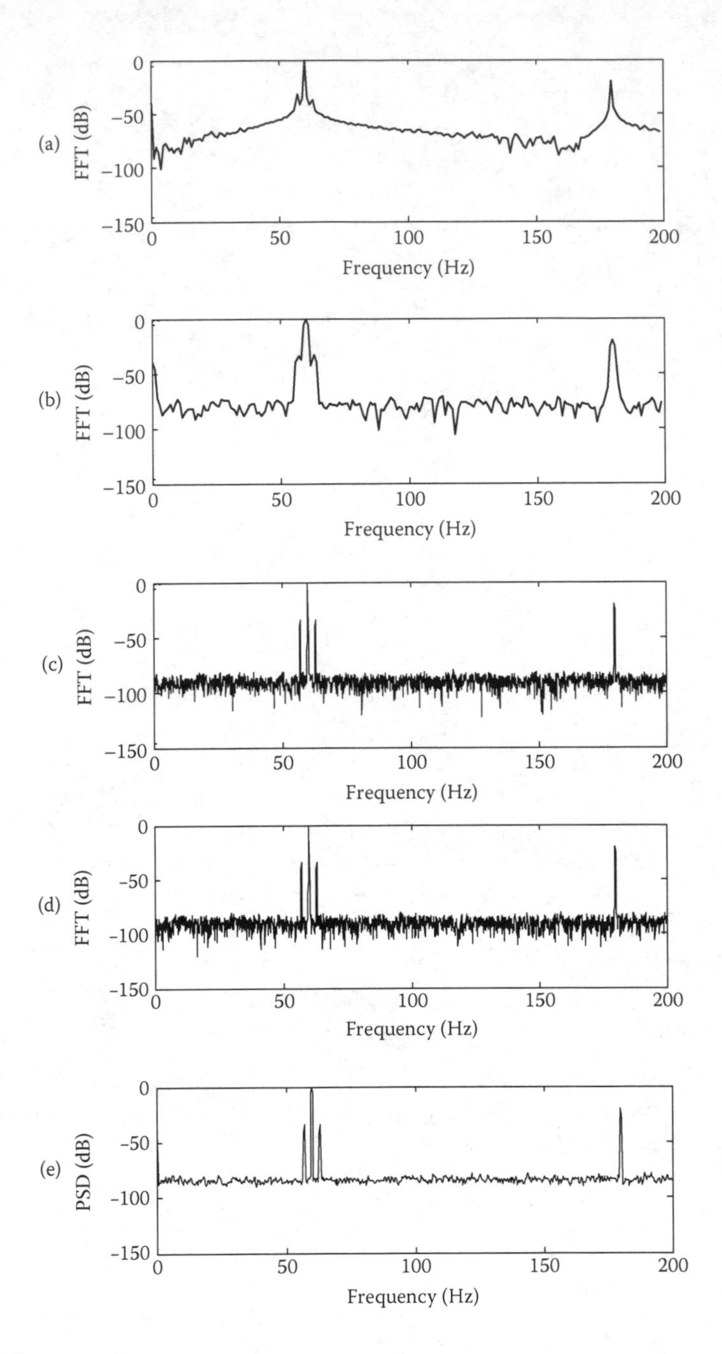

FIGURE 6.1
From top: FFT using rectangular window and 1 second of data, FFT using Hanning window and 1 second of data, FFT using rectangular window and 10 seconds of data, FFT using Hanning window and 10 seconds of data, PSD. All the plots have been normalized with respect to the 60 Hz component.

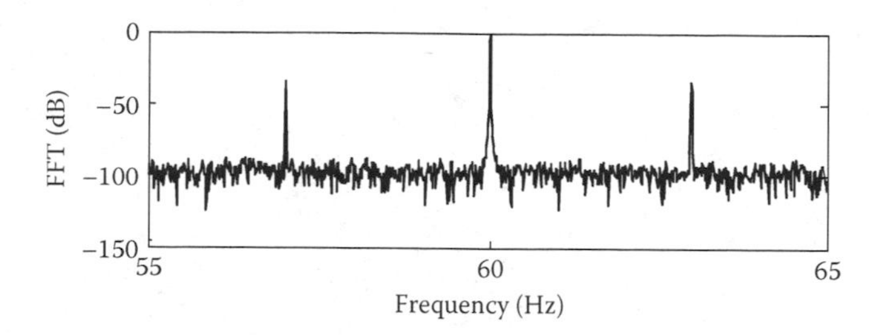

FIGURE 6.2
Zoomed spectra of signal in Example 6.3 around 60 Hz.

significantly reduced with this window. With a larger data set the effect of windowing is much less pronounced as can be seen in the next two plots. However, the frequency resolution has increased significantly with a larger data set. In the last plot the PSD spectrum shows significant reduction in noise with little sacrifice in resolution.

Sometimes it may be desirable to look closely at the narrow band of a spectrum, say around the 60 Hz frequency spectrum as described in Example 6.3. There exists a technique called *zoom FFT* [3,4] by which one can zoom on to the area of interest in a spectrum. This way one has to do far fewer numbers of FFT computations than would be required for the whole spectrum for a given resolution. To do this, the original collected data is shifted in frequency domain by multiplication with the sinusoid $e^{j\omega_1 t}$, where ω_1 is the lower limit of the band of interest. In the next step, the modulated signal is filtered with a low pass filter and then down sampled with a factor that essentially determines the "zoom." Figure 6.2 shows such a zoomed spectra of the signal described in Example 6.3 with a zoom or decimation factor of 10. The original signal was collected using a frequency of 3600 Hz for 100 seconds. Without zooming one has to do FFT of $3600 \times 100 = 360{,}000$ samples. With zooming by a factor of 10 it is reduced to 36,000. The resolution remains same as 0.01 Hz.

6.2.3 Parametric Power Spectrum Estimation

The nonparametric form of spectrum estimation techniques discussed in the previous section is fairly simple, well understood and easy to compute. However, they suffer from the fact that improved resolution of the spectrum would entail long data records. Thus estimation of fault signals for motors running perpetually under transient modes, such as hoists and winches will be difficult. Also, since finite-length data records are used, spectral leakage effects would be present. This would tend to mask weak signals present in the data, particularly in the vicinity of a strong signal.

Parametric or *model-based power spectrum estimation* methods eliminate the need for window functions and as a result the associated spectral leakage and frequency resolution problems [3,5–7]. Thus they hold promise for applications where short data records are available due to time-variant or transient phenomena.

The parametric techniques essentially assume that the data sequence whose spectrum has to be analyzed is the output of a linear system characterized by a rational transfer function in the discrete domain as

$$H(z) = \frac{B(z)}{A(z)} = \frac{\sum_{k=0}^{q} b_k z^{-k}}{1 + \sum_{k=1}^{p} a_k z^{-k}} \tag{6.12}$$

with

$$X(z) = H(z)C(z) \tag{6.13}$$

where $X(z)$ is the z transform of the output data sequence $x(n)$ to be analyzed and $C(z)$ the z transform of the corresponding input data sequence $c(n)$. Now if $c(n)$ is a zero mean, unity variance white noise sequence, it is easy to show that

$$|X(jw)|^2 = |H(jw)|^2 \tag{6.14}$$

It is very clear then that determining the sets $\{a_k\}, \{b_k\}$ in Equation (6.12) are enough to estimate the spectrum of $x(n)$.

Models such as those given by Equation (6.12) are generally known as *autoregressive-moving average* (ARMA) models. With $q = 0, b_0 = 1$ it is known as an *autoregressive* (AR) model. Setting $A(z)$=1 makes it a *moving average* (MA) model. The AR model is most widely used because of its simple form and suitability for representing spectra with narrow peaks. One of the most important aspects of AR models is the selection of the order p. If p is too low, the spectrum is very smooth. Too high values of p may end up producing spurious low level peaks in the spectrum.

Numerous techniques to obtain these models are available in literature. Yule–Walker, Burg, and unconstrained least squares are some examples. In special cases, when the signal components are sinusoids corrupted by additive white noise, eigenvalue-based techniques such as MUSIC and ROOTMUSIC have also been found useful.

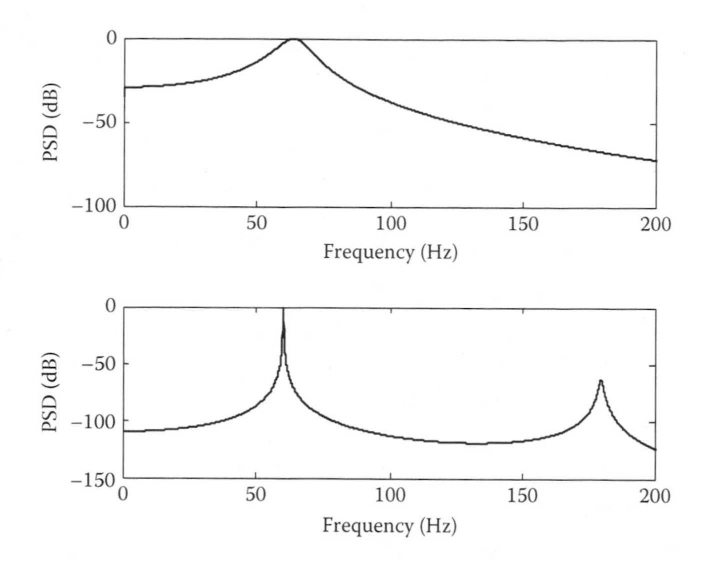

FIGURE 6.3
Spectrum of the signal described in Example 6.3 using Yule–Walker (top) and MUSIC (bottom) methods.

Figure 6.3 shows the spectral estimation carried out using the Yule–Walker and MUSIC method using MATLAB commands pyulear and pmusic. All 36,000 points were used. The order of the system was considered as four since there are four distinct sinusoidal signals. While the Yule–Walker method was just able to detect the 60 Hz component, the MUSIC was able to detect the 180 Hz signal as well. None were however able to detect the 57 and 63 Hz signals.

6.2.4 Power Spectrum Estimation Using Higher-Order Spectra (HOS)

Higher-order spectra (HOS) based spectral analysis has received some attention with regards to detection of very weak harmonics under low signal-to-noise ratio (SNR) conditions [6,8,9]. Recently it has been reported to have been used in building a tool called statistical motor analysis in real time (SMART), a PC-based software implementing fault diagnosis using HOS.

The average of the PSD is a second-order spectral measure since it essentially computes

$$P(\omega) = E[X(j\omega)X^{*}(j\omega)] \tag{6.15}$$

where $X(\omega)$ is same as the FT as described by Equation (6.5). $X^*(j\omega)$ is the complex conjugate of $X(j\omega)$ and $E[]$ is the statistical expectation or average. The same definition can be extended to obtain higher-order spectra such as

$$B(j\omega_1, j\omega_2) = E[X(j\omega_1)X(j\omega_2)X^*(j\omega_1 + j\omega_2)] \tag{6.16}$$

$$T(j\omega_1, j\omega_2, j\omega_3) = E[X(j\omega_1)X(j\omega_2)X(j\omega_3)X^*(j\omega_1 + j\omega_2 + j\omega_3)] \tag{6.17}$$

Equation (6.15) and Equation (6.16) are known as bispectrum and trispectrum, respectively. A close look at Equation (6.15) and Equation (6.16) suggest that if certain *frequencies along with their sums* are present, their presence can be detected very easily even with low SNR. The principle of HOS can be shown using the following example.

Example 6.4

Find the bispectrum of (a) $x(t) = \cos(\omega_a t)$ and (b) $x(t) = \cos(\omega_a t) + \cos(\omega_b t) + \cos\{(\omega_a + \omega_b)t\}$. Assume zero noise.

a. Following Example (6.2)

$$B(j\omega_1, j\omega_2) = \pi^3\left[\delta(\omega_1 - \omega_a) + \delta(\omega_1 + \omega_a)\right]\left[\delta(\omega_1 - \omega_b) + \delta(\omega_1 + \omega_b)\right]$$
$$\left[\delta(\omega_1 + \omega_2 - \omega_a) + \delta(\omega_1 + \omega_2 + \omega_a)\right] = 0.$$

This is because none of the impulses occur at the same frequency point.

b. In this case however,

$$B(j\omega_1, j\omega_2)$$
$$= \pi^3[\delta(\omega_1 - \omega_a) + \delta(\omega_1 + \omega_a) + \delta(\omega_1 - \omega_b) + \delta(\omega_1 + \omega_b) + \delta(\omega_1 - \omega_a - \omega_b)$$
$$+\delta(\omega_1 + \omega_a + \omega_b)]$$
$$[\delta(\omega_2 - \omega_a) + \delta(\omega_2 + \omega_a) + \delta(\omega_2 - \omega_b) + \delta(\omega_2 + \omega_b) + \delta(\omega_2 - \omega_a - \omega_b)$$
$$+ \delta(\omega_2 + \omega_a + \omega_b)]$$
$$\begin{bmatrix} \delta(\omega_1 + \omega_2 - \omega_a) + \delta(\omega_1 + \omega_2 + \omega_a) + \delta(\omega_1 + \omega_2 - \omega_b) + \delta(\omega_1 + \omega_2 + \omega_b) \\ + \delta(\omega_1 + \omega_2 - \omega_a - \omega_b) + \delta(\omega_1 + \omega_2 + \omega_a + \omega_b) \end{bmatrix}$$

Clearly here with $\omega_1 = \omega_a, \omega_2 = \omega_b$ or $\omega_1 = \omega_b, \omega_2 = \omega_a$, $B(j\omega_1, j\omega_2)$ is nonzero.

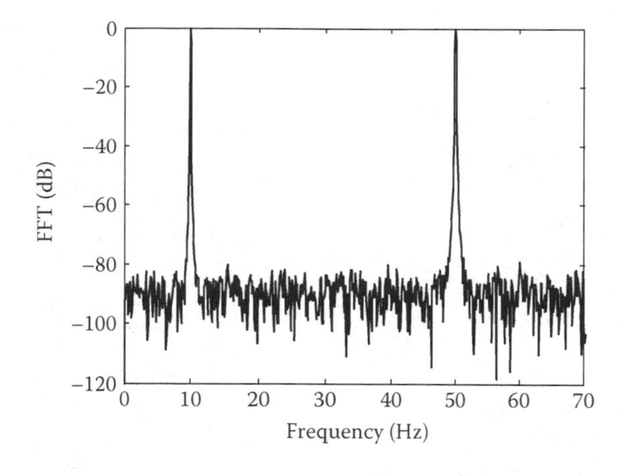

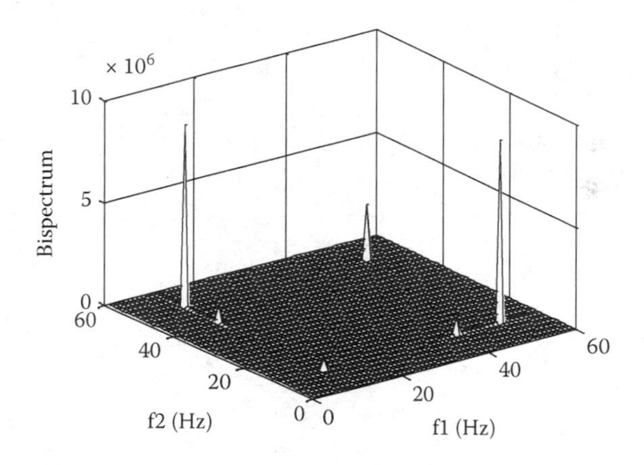

FIGURE 6.4
The FFT plot (top) and the bispectrum (bottom) plot of a signal with a very weak 60 Hz component.

The FFT (with Hanning window) and bispectrum of a signal with equal amplitude of $f_a = 10\text{Hz}$ and $f_b = 50\text{Hz}$ is given in Figure 6.4. None of them used any averaging technique. The signal also has a 60 Hz component whose amplitude is 0.01% of either $\omega_a$or ω_b. The SNR is about –20 dB. It is very clear that the 60 Hz component can be identified much better from the bispectrum plot. It is shown by two largest peaks located at the grid points $f_1 = 10\text{Hz}$, $f_2 = 50\text{Hz}$ and $f_1 = 10\text{Hz}$, $f_2 = 50\text{Hz}$. There are other minor peaks located at other grid points due to noise present in the signal. On the other hand, the 60 Hz signal is almost buried in the noise floor of the FFT output. It may be argued that the spectral quality of the FFT output can be improved by averaging. However it would mean increasing the computational overhead also.

6.2.5 Power Spectrum Estimation Using Swept Sine Measurements or Digital Frequency Locked Loop Technique (DFLL)

So far our attention was primarily focused on locating many spectral lines over a wide frequency range. However, many times if the frequency itself is roughly known or it varies only over a very narrow band under all operating conditions, then FFT analysis may not computationally be the best option.

The key to understanding the swept sine measurements or digital frequency locked loop technique (DFLL) [10,11] lies in the evaluation of the integrals $\int_0^{2\pi} \cos nx \cos mx\, dx$, $\int_0^{2\pi} \cos nx \sin mx\, dx$, and $\int_0^{2\pi} \sin nx \sin mx\, dx$. It is easy to show that they are all equal to π if $m=n$ and 0 if $m \neq n$. Essentially this means that if m denotes the frequency of interest in the signal then we can vary n over a narrow range to find the magnitude and location of m with great accuracy. Even the phase of the signal can be known. It is computed in the following way. If $f(t)$ is the signal, then the following are computed at regular interval $\Delta\omega$ in the range $\omega_1 \leq \omega \leq \omega_2$ in which the frequency of interest in $f(t)$ is expected to lie in

$$a = \int_0^T f(t) \cos \omega t\, dt \tag{6.18}$$

$$b = \int_0^T f(t) \sin \omega t\, dt \tag{6.19}$$

$$M = \sqrt{a^2 + b^2} \tag{6.20}$$

$$P = \tan^{-1} \frac{b}{a} \tag{6.21}$$

M designates the magnitude of the signal and P the phase. The location at which the peak of M occurs gives the frequency of interest. The frequency resolution is $\frac{1}{\Delta\omega}$. Normally, T spans over several cycles of the steady-state signal for improved detection. Also the products $f(t)cos\omega t$, $f(t)sin\omega t$ have to be suitably low pass filtered before the integration. Figure 6.5 shows the detection of the 60 Hz component in a signal that has equal amplitude 120 Hz signal and 0 mean, 0.3 standard deviation white noise. The SNR is about 8 dB. The frequency resolution is 0.1 Hz. One second of data was used. To get similar resolution using FFT would require data collected over 10 seconds.

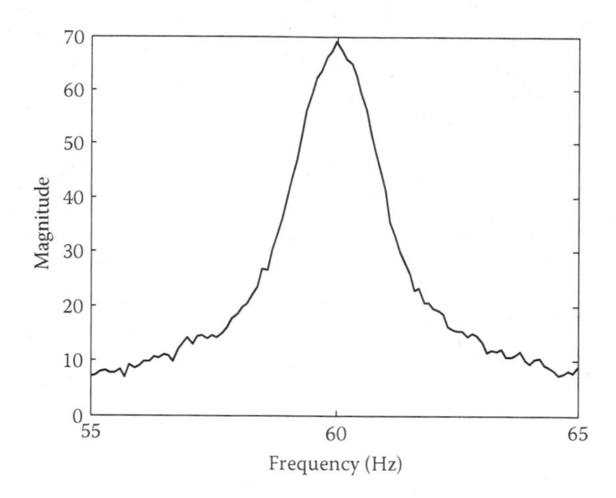

FIGURE 6.5
Detection of frequency signal using DFLL.

6.3 Diagnosis of Machine Faults Using Frequency-Domain-Based Techniques

In the rest of the chapter, the diagnosis of the four most common faults encountered in electric machines, namely, the bearing faults, the stator interturn faults, the broken rotor bar faults, and the eccentricity faults, using frequency-domain-based techniques will be discussed.

6.3.1 Detection of Motor Bearing Faults

As stated earlier, bearing faults happen to be the most common cause of electric machine failure in industry. Also, bearings faults have been recently classified as *single-point defects* that produce predictable frequencies and *generalized roughness* that do not. The two most common ways to determine single-point defects of bearings are by mechanical vibration and current signature analysis [12,13]. Of these the mechanical vibration signal analysis are most popular and will be discussed first. The current signature analysis of bearings is comparatively new and seems to be a function of the mechanical vibration signals. They will be discussed later.

6.3.1.1 Mechanical Vibration Frequency Analysis to Detect Bearing Faults

Most of the literature on fault detection of bearings deals with rolling-element bearings [14–17]. Most common among them are the ball bearings. They consist typically of six to twelve balls inserted in a perforated cage in

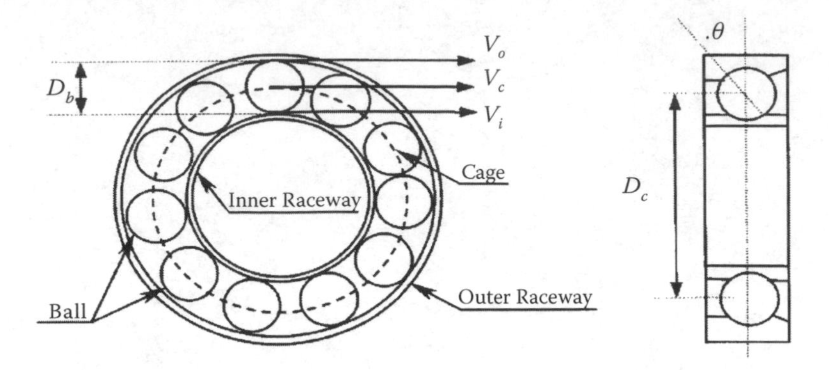

FIGURE 6.6
The different parts of a ball bearing (From Li B, M.-Y. Chow, Y. Tipusuwan, and James C. Hung, Neural-network-based motor rolling bearing fault diagnosis, IEEE Trans. on Industrial Electronics, pp. 1060–1069, vol. 47, no.5, Oct. 2000. With permission.)

the form of a ring. The cage ensures uniform spacing and prevents mutual contact. The cage with the balls is held by an outer ring known as outer raceway and an inner ring known as inner raceway. The balls are lubricated with grease. Other types of rolling element bearings use cylinders instead of balls. Sometimes the ends of the bearings are sealed. Very large electric motors use sleeve (fluid-film) bearings. Magnetic bearings are also possible.

The structure of the ball bearing is given in more detail in Figure 6.6. Let the cage, and outer and inner raceway velocities be V_c, V_o, and V_i respectively. These velocities essentially determine the different mechanical vibration frequencies associated with the cage, the ball, the outer raceway, the inner raceway, and the shaft. They are commonly known as the fundamental cage frequency (F_C), the ball rotational frequency (F_B), the ball pass outer raceway frequency (F_{BPO}), and the ball pass inner raceway frequency (F_{BPI}). All of them are a function of shaft rotational frequency (F_S). If D_b is the ball diameter and D_c is the bearing cage diameter, then following fundamental physics that relates angular velocity with linear velocity

$$F_C = \frac{V_c}{r_c} = \frac{V_o + V_i}{D_c} \tag{6.22}$$

where $V_c = (V_o + V_i)/2$ and $r_c = D_c/2$. Also due to the contact angle θ, only a part of the D_b will contribute toward the frequencies F_o (corresponding to V_o) and F_i (corresponding to V_i). Defining $r_i = r_c - (D_b \cos\theta/2)$, $r_o = r_c + \left(D_b \cos\theta/2\right)$, and with $V_o = F_o r_o$, $V_i = F_i r_i$, one could easily reduce Equation (6.22) to

$$F_C = \frac{1}{D_c}\left(F_i \frac{D_c - D_b \cos\theta}{2} + F_o \frac{D_c + D_b \cos\theta}{2}\right) \tag{6.23}$$

Similarly F_{BPI} and F_{BPO}, which indicates the rate at which the balls pass a point on the track of the inner and the outer raceway, respectively, can be expressed as the product of the number of balls and absolute difference in velocity between the cage and the inner or the outer raceway.

Thus, using Equation (6.23)

$$F_{BPI} = N_B |F_C - F_i| = \frac{N_B}{2} \left| (F_i - F_o) \left(1 + \frac{D_b \cos\theta}{D_c} \right) \right| \tag{6.24}$$

$$F_{BPO} = N_B |F_C - F_o| = \frac{N_B}{2} \left| (F_i - F_o) \left(1 - \frac{D_b \cos\theta}{D_c} \right) \right| \tag{6.25}$$

Finally, F_B, which indicates the rate at which it rotates around its own axis, can be calculated as

$$F_B = \left| (F_C - F_i) \frac{r_i}{r_b} \right| = \left| (F_i - F_o) \frac{r_o}{r_b} \right| = \frac{D_c}{2D_b} \left| (F_i - F_o) \left(1 - \frac{D_b^2 \cos^2\theta}{D_c^2} \right) \right| \tag{6.26}$$

Since in a motor the outer raceway is tightly fixed to the static end bells of a motor, $F_o = 0$. Similarly the inner raceway sits tightly on the rotor and rotates at the angular velocity F_S, and therefore $F_i = F_s$.

Thus Equations (6.23) to (6.26) can be written as

$$F_C = \frac{F_S}{2} \left(1 - \frac{D_b \cos\theta}{D_c} \right) \tag{6.27}$$

$$F_{BPI} = \frac{N_B}{2} F_S \left(1 + \frac{D_b \cos\theta}{D_c} \right) \tag{6.28}$$

$$F_{BPO} = \frac{N_B}{2} F_S \left(1 - \frac{D_b \cos\theta}{D_c} \right) \tag{6.29}$$

$$F_B = \frac{D_c}{2D_b} F_S \left(1 - \frac{D_b^2 \cos^2\theta}{D_c^2} \right) \tag{6.30}$$

For some types of bearings Equation (6.28) and Equation (6.29) can be approximated as

$$F_{BPI} = 0.4 N_B F_S \tag{6.31}$$

$$F_{BPO} = 0.6 N_B F_S \tag{6.32}$$

In case of a single-point bearing defect, only one of the four characteristic frequencies given by Equations (6.27) to (6.30) would show up. The collision between the bearing defects at the point of contact sets shockwaves that excite the natural resonance frequencies of machines. These frequencies act as carriers to the fault signature frequencies given by Equations (6.27) to (6.30), which could be treated as baseband signals. If f_c is the carrier frequency and f_b is the baseband signal, then components such as f_c, f_b, $f_c + f_b$, $f_c - f_b$ will be present. Since $f_c + f_b$, $f_c - f_b$ are produced by f_c, f_b their phases are also sum and difference of f_c and f_b respectively. This type of interaction is known as quadratic phase coupling (QPC) and is best detected by the bispectrum as given by Equation (6.15) or by bicoherence, a normalized form of bispectrum. However, since f_b is also present, f_c can be erroneously detected as a sum of f_b and $f_c - f_b$ if traditional bispectrum or bicoherence is used. Additionally, due to large mechanical damping at low frequencies, signals such as f_b can be significantly attenuated. Thus a modified bispectrum and bicoherence technique is proposed, where only the carrier, sum, and difference frequencies are included $(\omega_b = 2\pi f_b; \omega_c = 2\pi f_c)$

$$B(j\omega_c, j\omega_b) = E[X\{j(\omega_c + \omega_b)\}X\{j(\omega_c - \omega_b)\}X^*(j\omega_c)X^*(j\omega_c)] \tag{6.33}$$

$$|b(j\omega_c, j\omega_b)| = \frac{|B(j\omega_c, j\omega_b)|^2}{E\{|X(j\omega_c)X(j\omega_c|^2\}E[|X\{j(\omega_c + \omega_b)\}X\{j(\omega_c - \omega_b)\}|^2]} \tag{6.34}$$

It is easy to see from Equation (6.33) that zero phase angle is obtained when the carrier sum and difference frequencies are related, maximizing the expected value. Otherwise they are random, returning an expected value of zero. This technique has been able to clearly detect even incipient outer raceway faults, which could not be detected using standard power spectrum estimation of the mechanical vibration signal.

Detecting single-point faults in the inner raceway is more difficult because the fault moves in and out of the static load zone when the inner raceway is constantly moving. As a result, not only are the F_{BPI} frequencies modulated by the machine natural resonance frequencies but also the shaft rotational frequencies. Thus the fault frequencies occur in a group near the natural resonance frequency, each group containing several peaks separated by the shaft rotational frequency. The spacing from any peak in one group to another peak in another group can be given as

$$F_{SB} = \pm F_{BPI} + mF_S, m = 0, \pm 1, \pm 2... \tag{6.35}$$

The fault-finding formula is now modified from Equation (6.33) and Equation (6.34) to $(\omega = 2\pi f; \omega_{SB} = 2\pi f_{SB})$

$$B(j\omega) = E[X\{j(\omega + \omega_{SB})\}X\{j(\omega - \omega_{SB})\}X^*(j\omega)X^*(j\omega)] \tag{6.36}$$

$$|b(j\omega)| = \frac{|B(j\omega)|^2}{E\{|X(j\omega)X(j\omega|^2\}E[|X\{j(\omega + \omega_{SB})\}X\{j(\omega - \omega_{SB})\}|^2]} \tag{6.37}$$

The fault is now detected by counting the number of peaks for each value of m. It may be possible to extend this scheme for detection of the ball as well as cage defect.

However in the case of general roughness defect it is preferable to measure the root mean square (RMS) value of the mechanical vibration signal over a specified frequency range.

In general, the most appropriate measure to identify a fault varies from bearing to bearing [18]. Mechanical vibration for constant low load levels (below 50%) can be detected as the fault progressed from incipient to advanced stages. For constant larger load levels (above 50%) the transitions could be random. Variable load can have significant effect on the fault development process. However, increased level of RMS value of the mechanical vibration signal seems to be an accurate estimator to diagnose advanced fault level.

6.3.1.2 Line Current Frequency Analysis to Detect Bearing Faults

Characteristic mechanical vibration frequencies described by Equations (6.27) to (6.30) can be seen in the line current spectrum of the motor also due to secondary effects [19,20]. Mechanical vibration causes radial displacement between stator and rotor, which can be treated as a combination of rotating eccentricities moving in clockwise and anticlockwise direction. This leads to the following frequencies in line current:

$$f_{bng} = \left| f_e \pm m f_v \right| \tag{6.38}$$

where f_v is one of the characteristic vibration frequencies and f_e the supply frequency. For example, brinelling is akin to single-point defects on both outer and inner raceways. Hence both F_{BPI} - and F_{BPO} -related frequencies show increase in the mechanical vibration spectrum. However in the current spectrum, the results are much less encouraging, especially for F_{BPI} . This sentiment was echoed again by Obaid et al. [12].

Recently efforts were also made to detect generalized roughness faults of bearings. Since it was observed that from mechanical vibration signal the RMS value of the vibration signal increases at the broadband level, it was felt that parametric spectral analysis of line current is a better approach to diagnose these faults. The popular, all pole, AR model is chosen for this purpose. Each time stator current is sampled the AR spectrum is estimated and stored. For each new spectral estimate, the mean spectral deviation (MSD) is computed. The MSD is the mean of the difference from each point in the spectral magnitude of the base line spectrum to the current spectrum. The MSD is computed and recorded every time the stator current is sampled, and the change in MSD is used as the fault index. Choosing MSD alone as the fault index usually works well for low motor loads. However for higher motor loads, the average value of several MSD readings gave better results [20].

6.3.2 Detection of Stator Faults

The most important quality any scheme that detects stator faults must have is quickness of detection. Stator faults usually progress from incipient to a very advanced stage in a matter of seconds. Unless detected early enough, it might lead to fire, explosion, and even loss of personnel. Traditionally, stator faults are detected on-line by the negative sequence voltage or reduction of negative sequence impedance. However, voltage unbalance and machine asymmetries that also change the negative sequence current and impedance can cause misdiagnosis when faults involve only a few turns. While the machine asymmetries can be accounted for on a temporal basis, its robustness toward aging effects of a motor is yet to be vindicated. Also, stator fault will, by its very nature, create some unbalance that cannot be measured by measuring the terminal conditions. Thus, till the present date, very low turn fault detection has remained as a major challenge to researchers.

There exists a few turns fault detection scheme utilizing frequency-domain-based techniques. They are not as popular as the schemes mentioned in the previous paragraph. However, most of them are relatively new concepts and further research is required for their improvement. They will be discussed next.

6.3.2.1 Detection of Stator Faults Using External Flux Sensors

The earliest work on stator fault detection using external flux sensors was reported on by Penman et al. [21]. It was later experimentally proven by Penman et al. [22]. The basis of the stator fault detection using this approach lies in the fact that in an ideal machine, the axial flux of the machine is zero. In the presence of small machine inherent asymmetries they are still small. However, a stator fault causes large asymmetry and this produces components such as

$$f_{ss} = \left(k \pm n\frac{(1-s)}{p}\right)f \tag{6.39}$$

in the axial flux, where k is the order of the time harmonic, n the order of the shorted coil space harmonic, s the slip, p the number of pole pairs, and f the supply frequency. Using k =1 and n =1,2 one could compute these frequencies as 36.24, 48.12, 71.88, and 83.76 Hz. The last three of these were shown to increase under stator fault when tested on a 200 hp, 50 Hz, 8 pole slip-ring induction motor. A large coil with around 300 turns on a Plexiglas former was mounted concentrically around the shaft to detect the fault. Fault location was detected using four smaller symmetrically mounted coils of about 100 turns each on a plastic former and also mounted on the shaft.

Recently external flux sensors have been shown to diagnose stator faults even in variable speed drives using an 11 kW, 50 Hz, 4 pole squirrel-cage induction motor [23]. Some new frequency components as given by

$$f_{ss} = \left(\gamma \frac{R(1-s)}{p} \pm \nu \right) f \tag{6.40}$$

have been detected in the axial flux, where ν is the order of the time harmonic, R is the number of rotor bars, $\gamma = 0,1,2,3...$, the order of rotor space harmonic, s the slip, p the number of pole pairs, and f the supply frequency. Similar frequency components are well known to be present in case of eccentricity faults of induction motors. Therefore, confusion may arise as to what type of fault is being detected. Additionally, none of these schemes have been shown to be immune to voltage unbalance.

6.3.2.2 Detection of Stator Faults Using Line Current Harmonics

One of the earlier publications that discussed the line harmonic current increase due to stator faults also discussed components similar to Equation (6.39) and Equation (6.40) being found due to rotor faults in the line current [24]. According to Stavrou et al. [24], the stator current harmonics that are expected to vary due to stator inter-turn fault, and have their origins in stator current, are given by

$$f_{sc} = \left(j_{rt} R \frac{1-s}{p} \pm 2 j_{sa} \pm i_{st} \right) f \tag{6.41}$$

and those that have their origins in rotor current are given by

$$f_{rc} = \left((j_{rt} R \pm k) \frac{1-s}{p} \pm 2 j_{sa} \pm i_{rt} s \right) f \tag{6.42}$$

Here $j_{rt}, j_{sa}, i_{rt}, i_{st}, k$ are integers. The subscripts *sa,rt,st* are related to saturation, rotor and stator.

Interestingly, the third harmonic in the line current was one of the components that was shown to increase under fault in the study by Henao et al. [23] and was expected to show increase in the study by Stavrou et al. [24]. However, further investigation into the cause of this harmonic and related experimental results regarding its increase under stator fault have been completed [25–27]. Joksimovic and Penman showed that the negative sequence current interacts with the fundamental slip frequency current in the rotor conductors to produce torque pulsating at twice the line current frequency [25]. The consequent speed ripple caused flux density components at three times the line current frequency with respect to stator. This induced the third harmonic in line current. A more recent paper reports detection of third harmonic component in line current as a signature for stator fault [26]. It was attributed to the third harmonic present in the supply voltage and also to inherent machine asymmetry and voltage unbalance. While there is no doubt that the third harmonic voltages would manifest themselves in the line current of the machine, the degree to which these harmonics are normally present in line current are much larger than the voltages themselves [27]. It was further pointed out by Nandi that the fundamental frequency reverse rotating field (caused by fundamental frequency voltage unbalance and constructional asymmetry of the machine) interacts with the fundamental of the saturation-induced specific permeance function to produce the large third harmonic current, due to the presence of a matching pole pair associated with the third harmonic flux density component [27]. This can be clearly seen from the simulated plots in Figure 6.7.

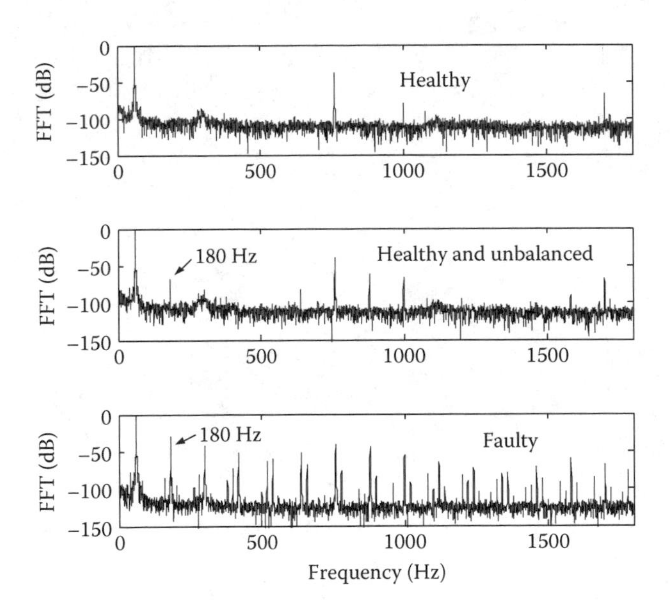

FIGURE 6.7
Simulated line current in the *b* phase of a 2.2 kW, 4 pole, 460V, 60 Hz induction motor with saturation when healthy (top), with 5% voltage unbalance (middle), and with 5 turns' fault in phase *a*. (From S. Nandi, "A detailed model of induction machines with saturation extendable for fault analysis," *IEEE Transactions on Industrial Application*, vol. 40, no. 5, pp. 1302–1309, September/October 2004. With permission.)

6.3.2.3 Detection of Stator Faults Using Terminal Voltage Harmonics at Switch-Off

The effect of voltage unbalance is naturally absent right from the moment it is switched off. However, due to the residual flux in the machine, current still flows in the rotor bars and also in the shorted coils in the stator. This fact has been utilized to detect the stator fault using rotor slot harmonics and later the more generic triplen-related harmonics [28]. According to Nandi and Toliyat [28], the voltage components induced by the shorted coil in the terminal voltage to detect is given by

$$f_v = [k(R/p) \pm 1] f_{off} \quad (6.43)$$

where k=1,2,3... and f_{off} is the frequency of the decaying stator voltage after switch-off and is proportional to the gradually diminishing speed of the induction machine. Figure 6.8 shows the simulated and Figure 6.9

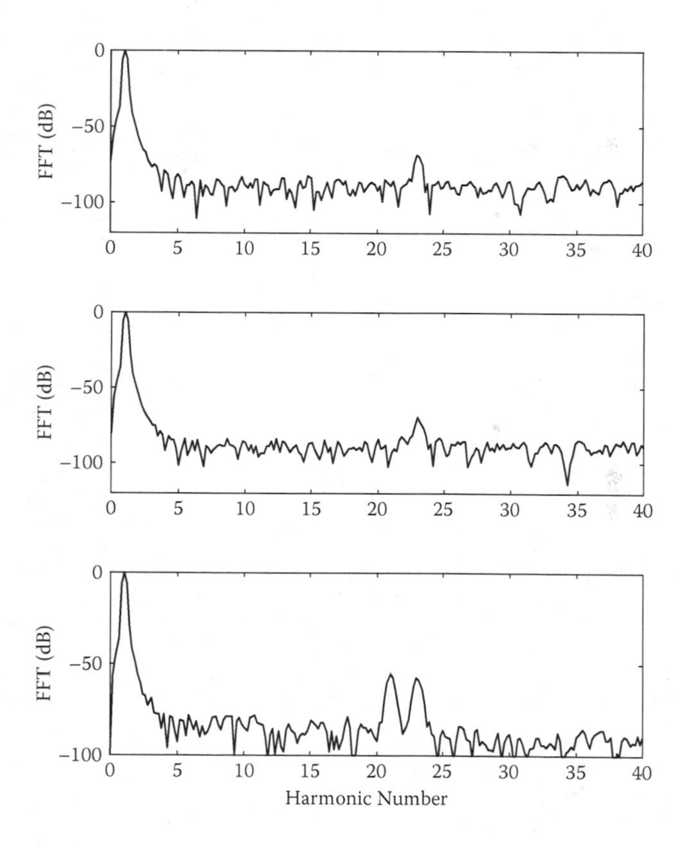

FIGURE 6.8
Simulated, normalized line voltage spectra of a 3 ph, 3 hp, 60 Hz, skewed, 44 bar, 4 pole induction motor under healthy (top), voltage unbalance (middle), and 5 turns fault in phase *a* (bottom) at switch-off.

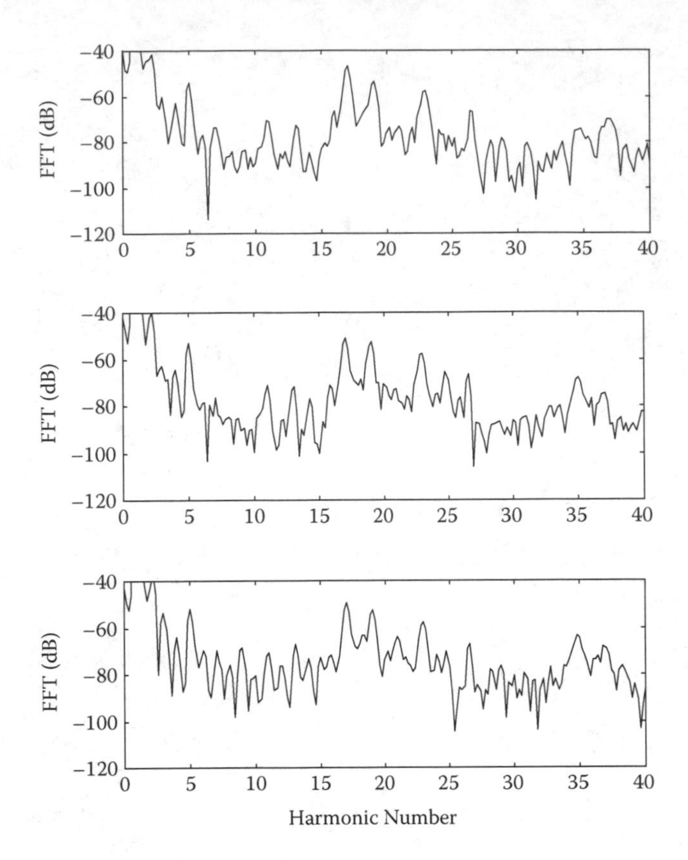

FIGURE 6.9
Experimental, normalized line voltage spectra of a 3 ph, 3 hp, 60 Hz, skewed, 44 bar, 4 pole induction motor under healthy (top), unbalanced (middle), and faulty with 5 turns' fault in phase *a* (bottom) at switch-off.

the experimental results respectively for a 2.2 kW, 44 bar, 4 pole 60 Hz machine. The 23rd harmonic was already present in the spectrum due to the stator winding and the flux pole pair matching. Hence the 21st harmonic was monitored. The faulty coil essentially produces all integral pole pairs and hence the 21st harmonic was induced in the voltage only under fault.

It was shown later by Nandi [29] that the odd triplen harmonics also showed increase with stator faults due to the presence of residual saturation given by

$$f_v = 3nf_{off}, n = 1, 3, 5... \tag{6.44}$$

where $n = 1,3,5...$. In this case also the odd triplen harmonics are induced as the matching pole pairs are present only under fault. These harmonics are produced in the flux by the interaction of air-gap magnetomotive force

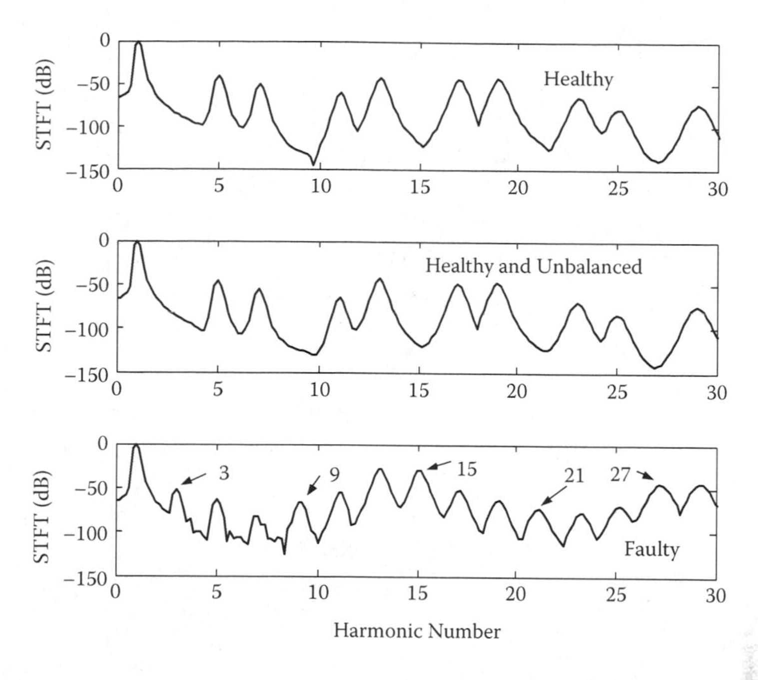

FIGURE 6.10
Simulated line voltage (*ab*) spectrum at switch-off of a saturated induction machine under healthy (top), unbalanced (middle), and with 5 turns fault in phase *a* (bottom).

(MMF) and saturation-related permeance components. Some simulation and experimental results are provided in Figure 6.10 and Figure 6.11. Detailed experimental results showed that faults with three turns and above under no load and two turns and above under full load could be unambiguously detected using this scheme.

Similar harmonics show up in case of reluctance synchronous motors (RSMs) also [30]. Figure 6.12 shows the results for a 1.5 hp, 460V RSM.

6.3.2.4 Detection of Stator Faults Using Field Current and Rotor Search Coil Harmonics in Synchronous Machines

Recently it was observed that due to the inherent asymmetry present in the field winding of the synchronous machines, harmonics of the form

$$f_r = [k(1/p) \pm 1]f, k = 1,2,3... \tag{6.45}$$

get induced in the field current. However, out of these only those harmonics not given by

$$f_n = [n \pm 1]f, n = 1,5,7,11,13... \tag{6.46}$$

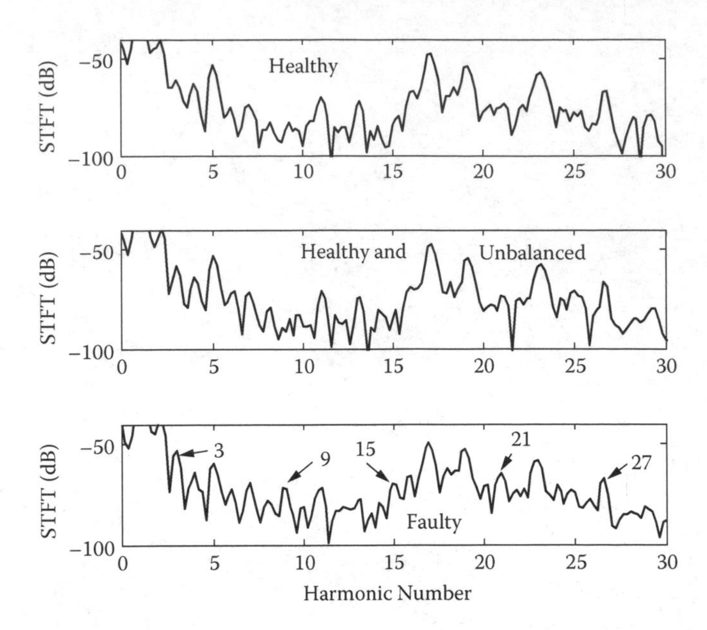

FIGURE 6.11
Experimental line voltage (*ab*) spectrum at switch-off of a saturated induction machine under healthy (top), unbalanced (middle), and with 5 turns fault in phase *a* (bottom).

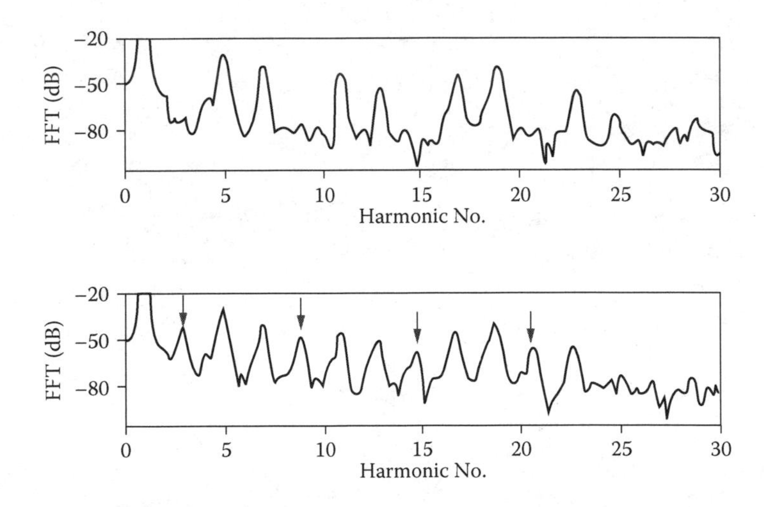

FIGURE 6.12
Experimental line voltage (*ab*) spectrum at switch-off of a RSM under unbalance (top) and with 5 turns fault in phase *a* (bottom).

can be considered, since harmonics given by Equation (6.46) can appear under balanced voltage (obtained with negative sign in Equation 6.46) as well as unbalanced voltage (obtained with positive sign in Equation 6.46) conditions. Since ideally the field coils can accept harmonics in the flux with only a certain number of pole pairs, the increase of the fault signature harmonics did not show sufficient increase for a low number of faulty turns and certain operating conditions. This is mostly due to the fact that these harmonics were induced due to inherent asymmetry of the field winding. However the rotor search coil, by construction, was capable of accepting harmonics of any integral pole pair number from the field current. Thus harmonics induced in the search coil can be used to detect even one turn fault under any operating condition [31]. Illustrative simulation and experimental results are shown in Figure 6.13 and Figure 6.14. HB

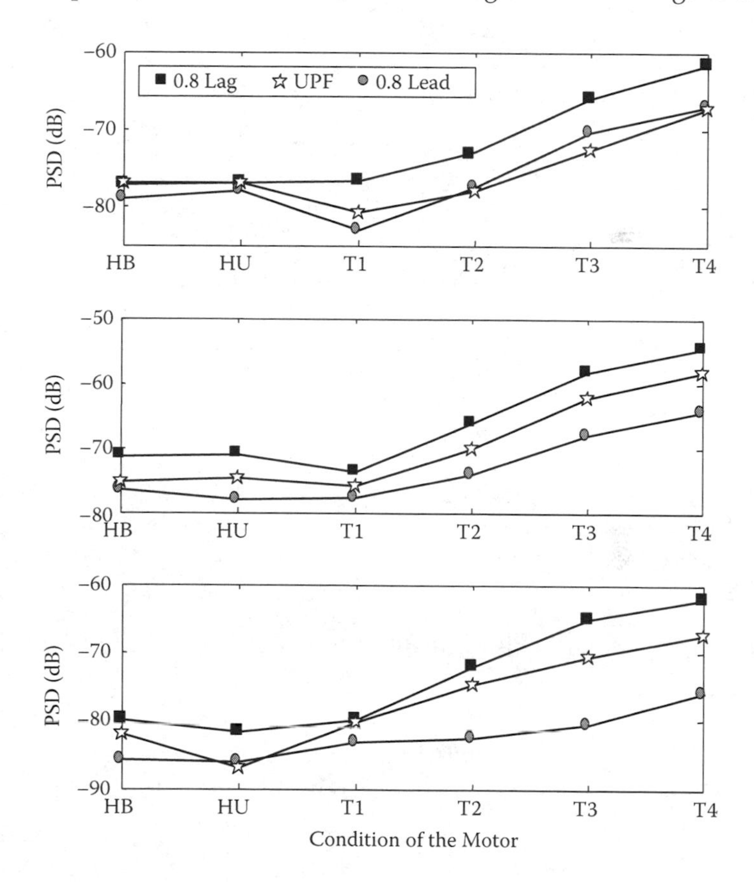

FIGURE 6.13
The 150 Hz component in the field current of a 2 kW, 4 pole, 60 Hz, 208V, synchronous motor, under no-load (top), half-load (middle), and full-load (bottom), 0.8 lagging power factor condition (experimental). 150 Hz can be obtained by using $k=3, p=2, f=60$ and the positive sign before 1 in Equation (6.45).

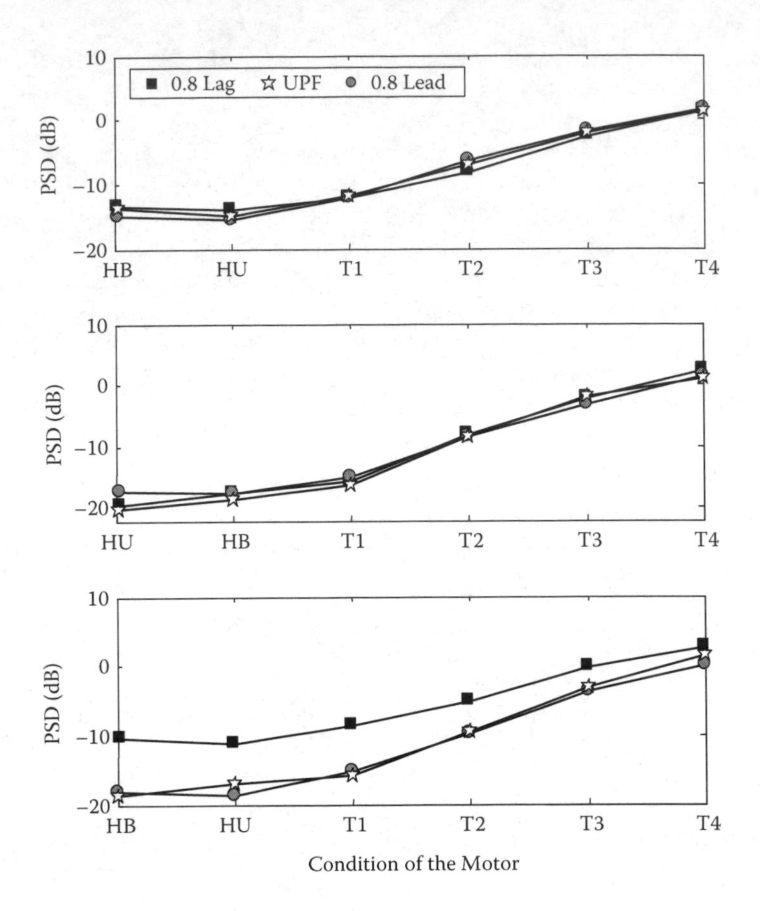

FIGURE 6.14
The 90Hz component in the rotor search-coil of a 2 kW, 4 pole, 60 Hz, 208V, synchronous motor, under no-load (top), half-load (middle) and full-load (bottom), 0.8 lagging power factor condition (experimental). 90 Hz can be obtained by using $k=3, p=2, f=60$ and the negative sign before 1 in Equation (6.45).

implies healthy balance, HU is healthy unbalance, and T1–T4 implies one turn to four turns short.

6.3.2.5 Detection of Stator Faults Using Rotor Current and Search Coil Voltages Harmonics in Wound Rotor Induction Machines

During an inter-turn fault the stator has a shorted loop (can thus be treated as a single-phase winding) carrying current at supply frequency that generates two counter-rotating MMF waves [32]. The MMF produced by the asymmetric stator carrying three-phase balanced voltage can be given as

$$F_{sa} = A_{sa}\cos(k\varphi \pm \omega_1 t + \gamma_1) \tag{6.47}$$

where $k = 1, 2, 3\ldots$ corresponds to space harmonic poles. Considering the specific permeance function (P_0) the flux density produced by this MMF, with respect to stator, can be given as

$$B_{sa} = A_{sa} P_0 \cos(k\varphi \pm \omega_1 t + \gamma_1) \tag{6.48}$$

With respect to rotor, this flux density can be given as

$$B_{ra} = A_{sa} P_0 \cos(k\varphi' + k\omega t \pm \omega_1 t + \gamma_1) \tag{6.49}$$

Now substituting $\omega = \frac{(1-s)\omega_1}{p}$ in Equation (6.49), we can have

$$B_{ra} = A_{sa} P_0 \cos\left(k\varphi' + \left\{\frac{k}{p}(1-s) \pm 1\right\}\omega_1 t + \gamma_1\right) \tag{6.50}$$

The term associated with t in Equation (6.50) gives the frequency component f_r that can used for detection as

$$f_r = \left\{\frac{k}{p}(1-s) \pm 1\right\} f_1 \tag{6.51}$$

For example, the frequencies that will be induced in the rotor circuit due to a fault in stator winding when a doubly fed induction generator (DFIG) is running at $s = 0.25$, $f_1 = 60$ Hz, $p = 2$ and different values of k are expressed in Table 6.1 using Equation (6.51). As seen from Table 6.1, several frequencies can be induced as a result of the fault. Unfortunately, many of the components given by Table 6.1 can be present even under healthy conditions and hence cannot be treated as reliable indicators of the fault.

Hence a detailed simulation study was conducted and compared with experimental results. Some of the very prominent components were 82.5 Hz for $k = 1$ and 127.5 Hz for $k = 3$, which arises due to asymmetry of the machine as can be seen from the simulated plots in Figure 6.15. Also the components

TABLE 6.1

Stator Fault Frequencies Induced in Rotor, $s = 0.25$

k	1	2	3	4	5	6	7	8	9
f_r (Hz)	82.5	105	127.5	150	172.5	195	217.5	240	262.5
	37.5	15	7.5	30	52.5	75	97.5	120	142.5
k	10	11	12	13	14	15	16	17	18
f_r (Hz)	285	307.5	330	352.5	375	397.5	420	442.5	465
	165	187.5	210	232.5	255	277.5	300	322.5	345

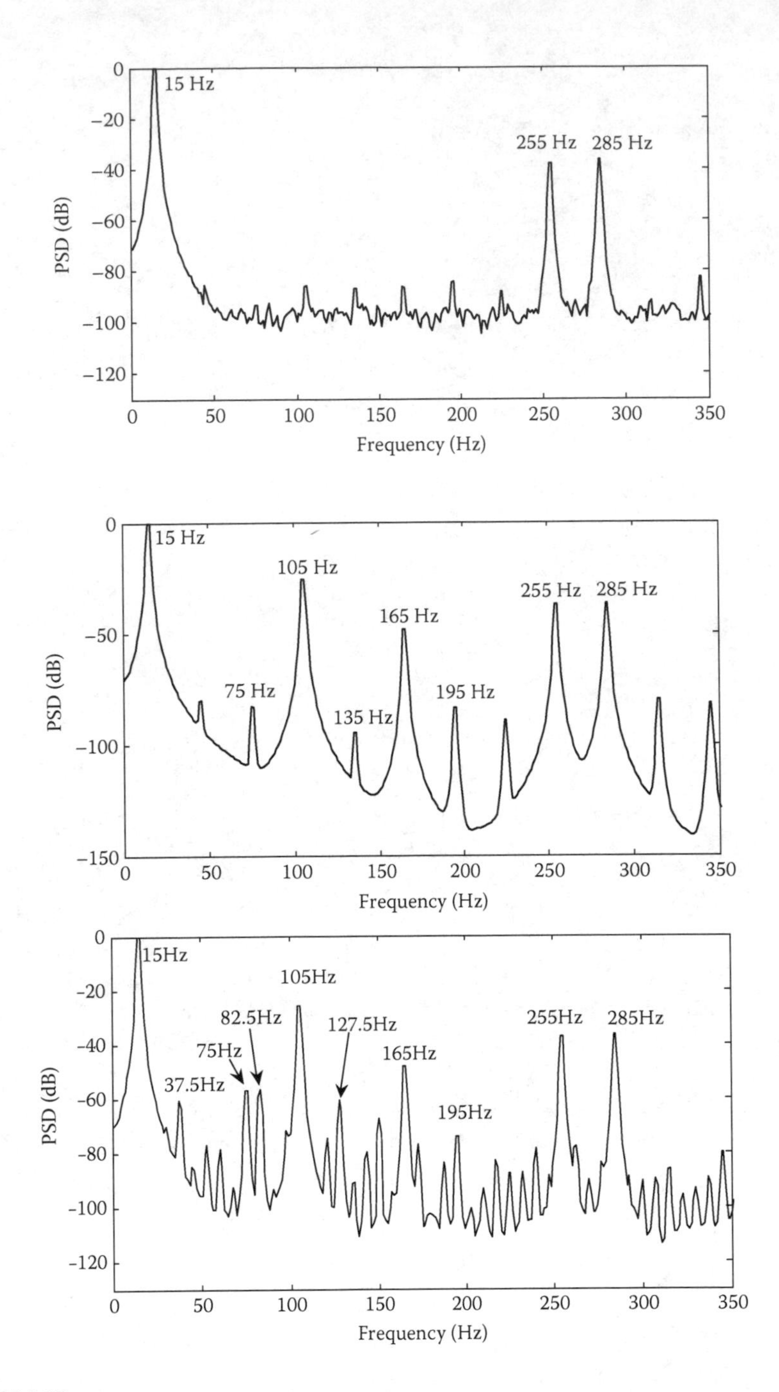

FIGURE 6.15
PSD of a simulated DFIG connected to balanced load with symmetrical rotor winding (top), symmetrical rotor winding subjected to 4-turn fault (middle), and asymmetrical rotor winding (1 reduced turn in one phase) subjected to 4-turn fault (bottom).

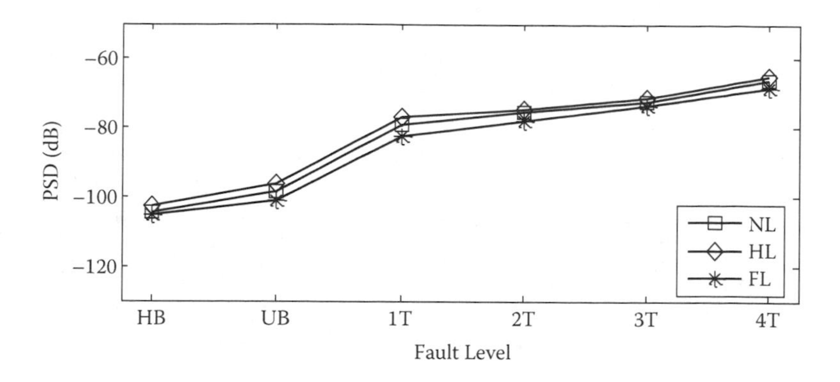

FIGURE 6.16
Simulation variation of fault spectra of rotor current frequency 127.5 Hz (k = 3) for slip 0.25 under no-load, half-load, and full-load, with varying fault severity. HB implies healthy balanced load, UB implies unbalanced load, 1T–4T implies fault levels from 1–4 turns.

related to k = 3 showed better promise as triplen-space-related harmonics seem more affected by asymmetry.

The model with one turn rotor asymmetry was further explored for unbalanced load and different fault levels. As can be seen from Figure 6.16, unbalanced load (10% on stator A phase) does not affect the result to a great extent. Also, the fault signature increased in proportion with the number of faulted turns.

The experimental results (Table 6.2) showed more consistent results for detection when rotor line current space vector was used rather than individual line current. The current space vector actually gave comparable results or even better results (at higher slip) with the rotor search coil.

The detection scheme was implemented on-line using the scheme shown in Figure 6.17. It worked quite reliably even down to two turns fault level, which can be detected within approximately 2 seconds (includes tripping signal to circuit breaker) (Figure 6.18). The scheme worked even under transient condition (Figure 6.19) quite reliably.

TABLE 6.2

Comparison of Signal-to-Noise Ratio (Given by the Difference between the Faulty and the Balanced Healthy Signature) for Fault Signature Frequency Component Power Level for Different Severity of Fault under Full-Load Condition

Slip	0.25	0.25	0.25	0.25	0.44	0.44	0.44	0.44
Number of Turns Faulty	1T	2T	3T	4T	1T	2T	3T	4T
Search Coil Voltage	1.73	9.05	14.5	18.02	0.03	0.52	6.92	9.29
Rotor Phase Current	–4.47	–2.68	3.46	8.07	–4.93	–3.59	1.29	6.58
Rotor Current Vector	3.81	3.70	10.92	14.70	1.18	1.75	5.28	11.52

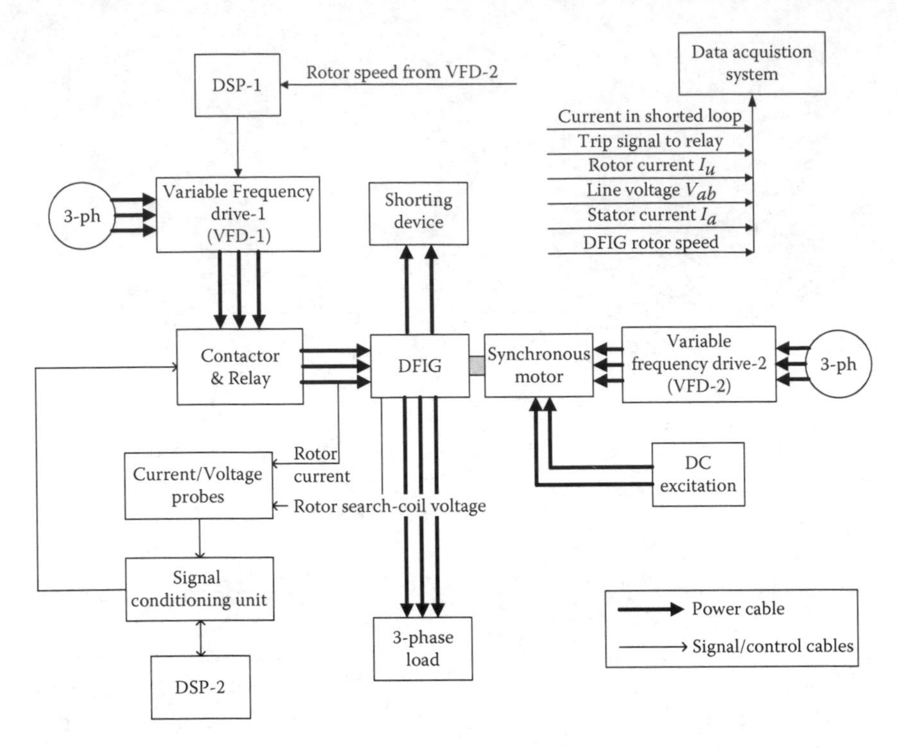

FIGURE 6.17
Schematic of experimental setup used to determine DFIG behavior under varying load, speed, fault severity, fault detection, and tripping.

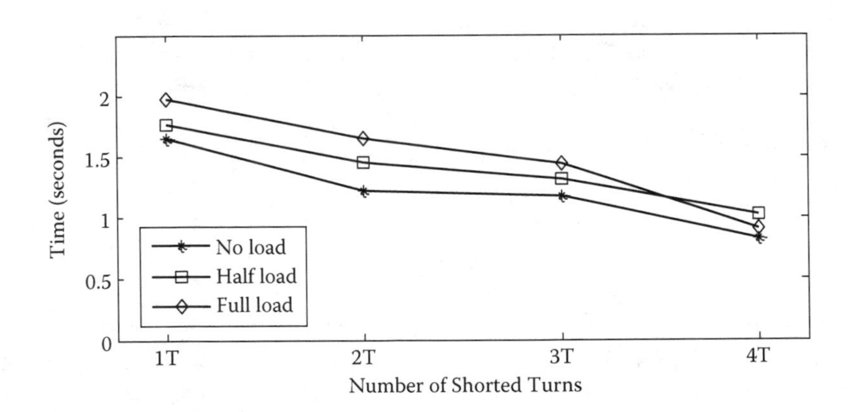

FIGURE 6.18
Typical time of operation of the DSP-based fault detection device when using search coil voltage signature analysis. DFIG operating at slip = 0.25 at different loads.

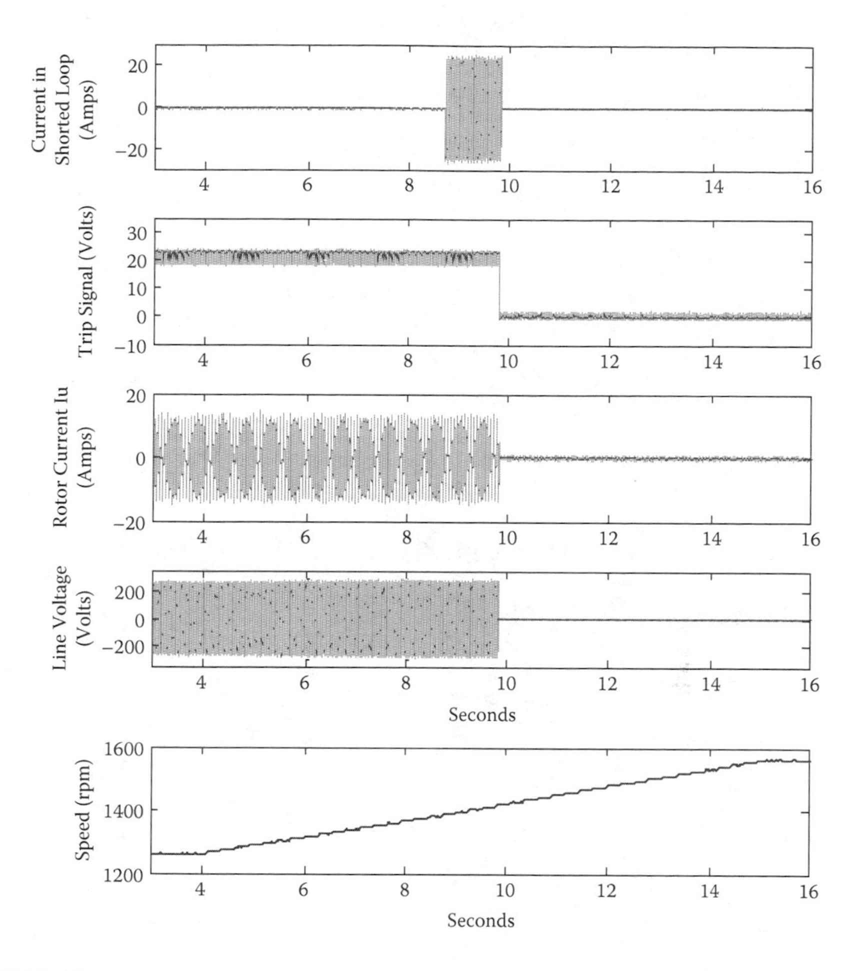

FIGURE 6.19
Various signals of the fault detection scheme, DFIG operating at half-load during a speed change, fault severity is 2-turn, fault detection time is 1300 msec.

6.3.3 Detection of Rotor Faults

Unlike a stator inter-turn fault, the rotor bar fault is an *open circuit* fault. Also, unlike stator inter-turn fault, it often does not lead to a catastrophe within a short period of time. Either rotor bars or end-rings may be open circuited. However, since the bars are typically not insulated, bar breakage at the initial stages may not be detectable due to the presence of interbar currents [33]. Also this type of fault can be detected only under loaded condition, since under no-load the rotor current is almost zero. Although many techniques to detect these faults exist, unlike stator inter-turn faults, detection of signature frequency components in the line current is the most common way to detect these faults. Some of the different frequency-domain-based methods are discussed next.

6.3.3.1 Detection of Rotor Faults in Stator Line Current, Speed, Torque, and Power

When a rotor bar is broken, there is an increase in the current distribution in the two bars adjacent to the broken rotor bar [34]. This can be deemed as current flowing in a single-phase winding, and therefore the double revolving field theory used for the analysis of single-phase induction motors can be applied. The anomalous MMF produced by these two bars with a rotor bar broken between them can be expressed in rotor coordinates as

$$F_s = F_m \cos(nx' \pm s\omega t) \tag{6.52}$$

where x' is the space angle with respect to rotor, ω is the supply frequency in rad./sec., s is the slip, $n = 1, 2, 3\ldots.$

MMF components as described by Equation (6.52) will induce voltages in the stator winding of a regular three-phase motor given by

$$v_s = v_m \cos\left(nx + \left(n\left(\frac{1-s}{p}\right) \pm s\right)\omega t\right) \tag{6.53}$$

only for n=p,5p,7p,11p,13p... as they alone can match the stator pole pairs. x' is the space angle with respect to the stator and $x = x' + (\frac{1-s}{p})\omega$. Therefore, generalized components in the current spectrum can be given as

$$f_s = \big(k(1-s) \pm s\big) f, k = 1,5,7,11,13... \tag{6.54}$$

with k=1 and the negative sign before s, the oft-quoted (1–2s)f component can be found.

The production of the other oft-quoted (1+2s)f is more subtle is nature. The sf component of the current anomaly in the rotor interacts with the air-gap flux and produces a 2sf component in the torque and hence in the speed. This induces a phase modulation of ±2sf in the stator flux that produces both (1–2s)f and (1+2s)f components in the stator current. This phenomenon gives rise to a sequence of additional current components at frequencies given by

$$f_b = (1 \pm 2ms) f, m = 1,2,3.... \tag{6.55}$$

Since the speed ripple is a secondary effect, (1–2s)f produced this way will affect the spectrum much less than that produced via the method described

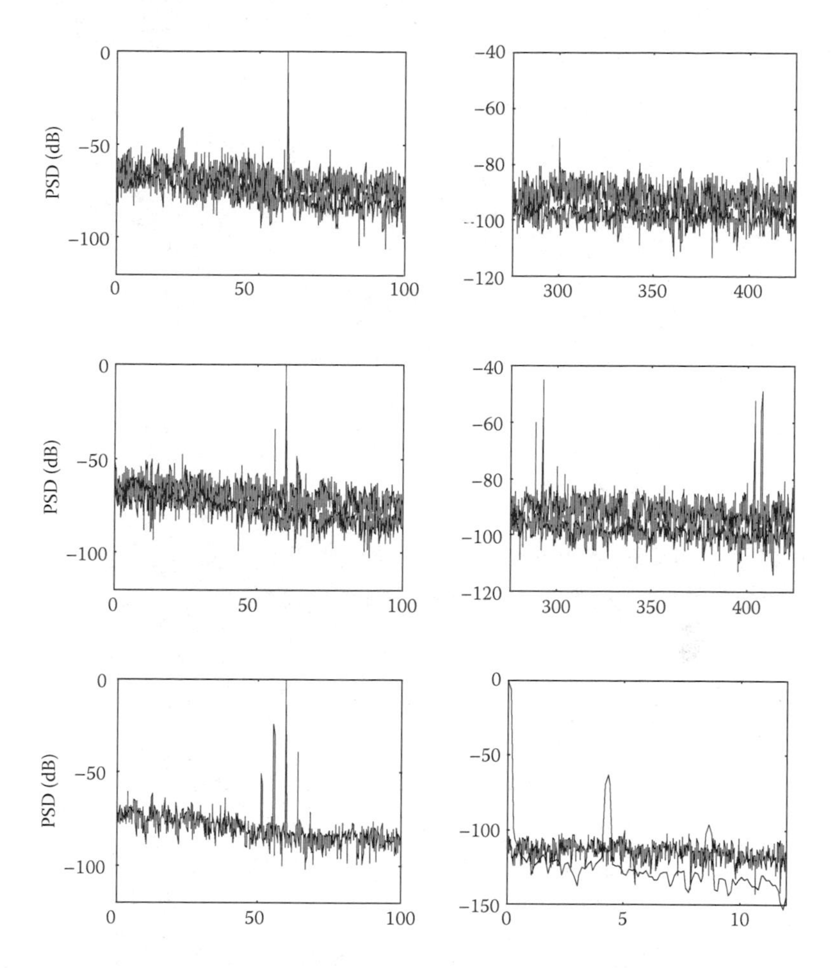

FIGURE 6.20
Simulated, normalized plots of phase *a* current spectra for the healthy (top row); with two cracked bars (middle row) around fundamental (left) and 5th and 7th harmonic components of line current (right); with two cracked end-rings (last row) around fundamental component of current (left) and speed. (From S. Nandi, "Fault analysis for condition monitoring of induction motors," PhD dissertation, Texas A&M University, May 2000. With permission.)

earlier. However since the $(1+2s)f$ component is produced by the speed ripple effect alone, inertia of the motor drive affects this signal much more.

Figure 6.20 [35] shows the simulated spectra of a 44-bar machine that has developed cracks in the bar or end-rings. Both current and speed spectra show the characteristic fault signatures. It also clearly shows that a broken end-ring is a more severe fault. However, in actual machines the signals may not be so distinctive because of interbar current and inherent bar-to-bar asymmetry that inherently produces the sideband components. Figure 6.21 [35] shows such an experimental plot for a healthy machine and one to four

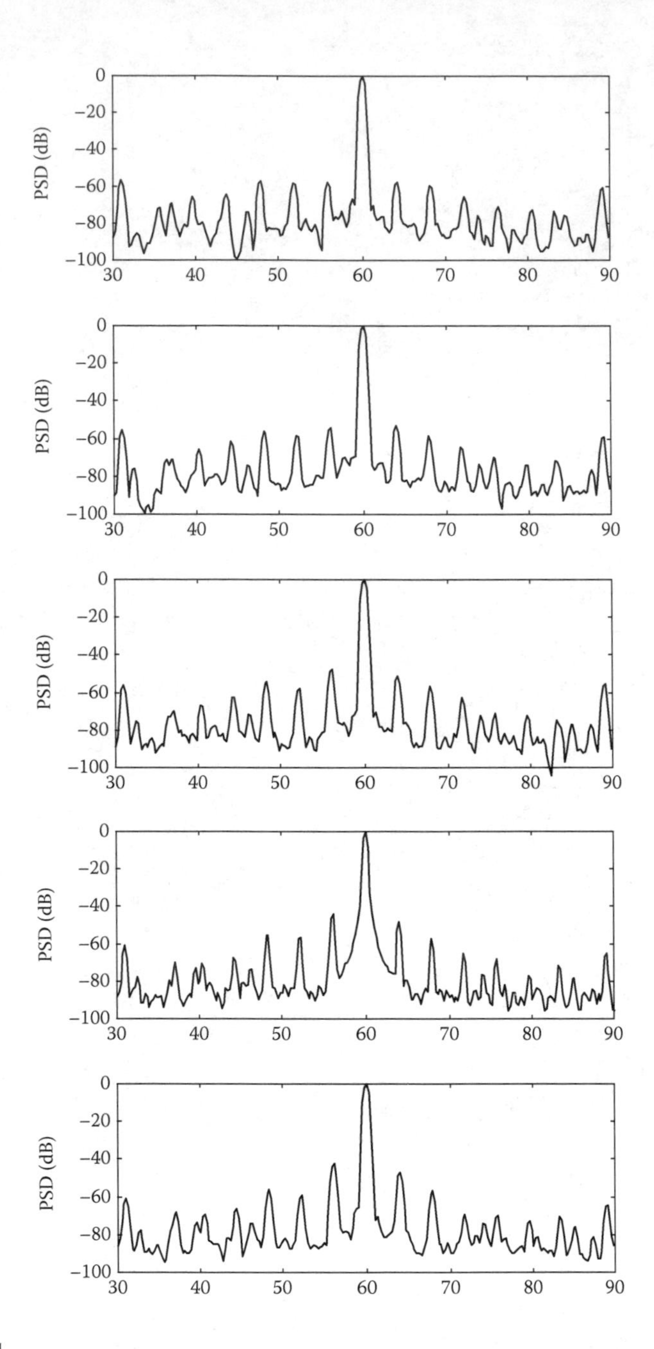

FIGURE 6.21
Experimental plots of normalized phase *a* current spectra of healthy machine (top) and with one to four rotor bars broken (next four plots). The $1 \pm 2sf$ components are right next to the 60 Hz fundamental. (From S. Nandi, "Fault analysis for condition monitoring of induction motors," PhD dissertation, Texas A&M University, May 2000. With permission.)

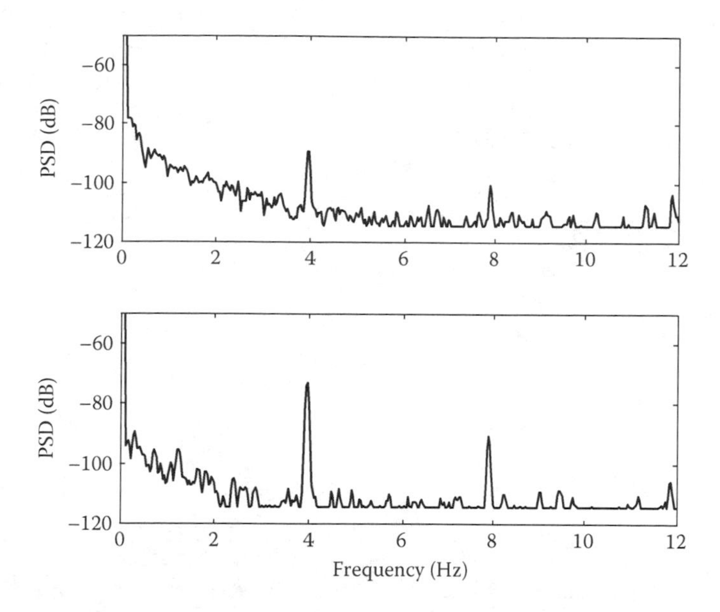

FIGURE 6.22
Experimental plots of normalized speed spectra of healthy machine (top) and with four rotor bars broken (bottom). (From S. Nandi, "Fault analysis for condition monitoring of induction motors," PhD dissertation, Texas A&M University, May 2000. With permission.)

rotor bars broken. The experimental speed spectra of the healthy and the four broken rotor bar machines are shown in Figure 6.22 [35]. The broken rotor bar machine is shown in Figure 6.23 [35]. The bars were broken by drilling holes. Due to high inertia of the motor-load system, the $(1+2s)f$ component shows little change in Figure 6.21 [35] in spite of large speed ripple.

FIGURE 6.23
Experimental machine. (From S. Nandi, "Fault analysis for condition monitoring of induction motors," PhD dissertation, Texas A&M University, May 2000. With permission.)

According to Filippetti et al., the sum of the values of the $(1-2s)f$ and $(1+2s)f$ components would give a better indication of load severity [36]. It has been proven and shown conclusively that the $2sf$ component in the motor instantaneous power also can be used to detect rotor bar faults [37].

6.3.3.2 Detection of Rotor Faults in External and Internal Search Coil

Rotor bar faults can also be detected by the presence of components such as

$$f_{sc} = ((1-s)/p \pm s)f \tag{6.56}$$

in external search coils. They can be shown to be present from Equation (6.53) for n=1. They were predicted for the first time by Kliman et al. [34], and later shown conclusively by Elkasabgy et al. [38]. Elkasabgy et al. also used internal search coils on the stator tooth tip as well as on the yoke. The yoke coil showed stronger results for the same number of broken rotor bars. Detection of these faults is also possible by spectral analysis of shaft flux [21] or more generally axial leakage flux (mentioned earlier with reference to stator fault detection), which is monitored by using an external search coil wound around the shaft of a machine. The components to look for are given by Equation (6.55). Interbar current induced axial flux in the presence of broken rotor bar faults can also be measured using shaft-mounted search coils at frequencies given by $\pm ksf, k =$ 1,3,5...[39]. The coil has to be mounted near the end where the rotor bars have broken.

6.3.3.3 Detection of Rotor Faults Using Terminal Voltage Harmonics at Switch-Off

Like stator faults, rotor faults also have been reported to be detectable using odd harmonics in the motor terminal voltage at motor switch-off [40]. However, this technique suffers from the fact that, for good detection, the broken rotor bars should be at the vicinity of peak of the current at the switch-off instant. Otherwise the fault may not be detected. Both simulation and experimental results are shown for a 3 hp, 60 Hz, 4 pole, 460V squirrel-cage induction motor in Figure 6.24.

6.3.3.4 Detection of Rotor Faults at Start-Up

In many applications it is very difficult to get prolonged periods of steady-state operations to perform reliable FFT or PSD. Even when they are, mechanical vibration caused by nonuniform air-gap may result in a very noisy current spectrum. A case in point is an electric or hybrid vehicle running in a city with frequent starts and stops. Zero-speed conditions were employed to overcome the aforementioned problems in detecting rotor faults

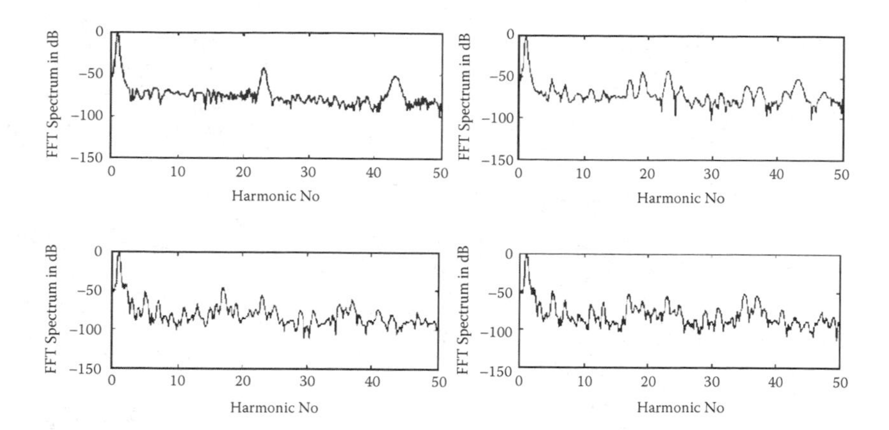

FIGURE 6.24
Top row: Simulated plot of healthy machine (left) and with four broken rotor bars (right). Bottom row: Experimental plot of healthy machine (left) and with four broken bars (right). In both simulation and experiment the 35th and the 37th harmonic showed substantial increase. (From J. Milimonfared et al., "A novel approach for broken rotor bar detection in cage induction motors," *IEEE Transactions on Industrial Applications*, vol. 35, no. 5, pp. 1000–1006, September/October 19. With permission.)

[41]. If Equation (6.55) is recomputed with m=1,s=1, it can be easily seen that detectable frequency with the negative sign

$$f_b = -f \tag{6.57}$$

The negative sequence of this current waveform can be easily located by computing the PSD of the line current space vector. The results can be obtained in real-time using a DSP.

6.3.3.5 Detection of Rotor Faults in Presence of Interbar Current Using Axial Vibration Signals

Analysis in [42] shows that axial vibration is increased because of increased interbar current in the presence of broken rotor bars [42]. The following frequencies are is shown to increase:

$$f_{v1} = \{(q_b - q_a) + s(q_a - q_b)\}f \tag{6.58}$$

$$f_{v2} = \{(2 - q_a - q_b) + s(q_a + q_b)\}f \tag{6.59}$$

The following modulated components given by

$$f_m = f_{v1} \pm f_c, f_{v2} \pm f_c \tag{6.60}$$

also showed increase. Here q_a, q_b are the standard space harmonics (1, 5, 7, 11, 13, …) of an induction machine, f is the supply frequency, f_c is the rotational frequency in hertz. With $q_a = 1, q_b = 7, f_c = 49Hz$, for a 2 pole, 50 Hz machine, $f_{v1} = 294Hz$ with s=0.02 and the modulation frequencies are 245 and 353 Hz.

6.3.4 Detection of Eccentricity Faults

Eccentricity faults are related to deformation of air-gaps of an electric machine. Hence it essentially affects the flux density of the motor or generator. Flux density affects currents indirectly and the mechanical vibration signals directly. Hence mechanical vibration signals still remain a very dependable means of detecting eccentricity. However, in a large plant with many other types of machinery, it may be difficult to pick up the vibration signals. Unfortunately, most of the time both current and vibration spectrum detection require the knowledge of the rotor or stator slots (does not appear on a motor nameplate) to detect the fault-related spectrum. The problem arises when, for some reason, this data is not available. Usually eccentricity is a major issue only with large machines, and the manufacturers will usually have this data readily available for such large machines. Nowadays even for small motors, such information is sometimes available on the Web.

6.3.4.1 Detection of Eccentricity Faults Using Line Current Signal Spectra

Flux density in the air-gap of any machine is due to the interaction of machine MMF and permeance. While the MMF is the function of time and space harmonics, the permeance is the function of rotor and stator slots and eccentricity. It can be shown, following that the high stator current components arising out of the static, dynamic, or a mixture of both [43–45] can be given as

$$f_{ecc1} = \left[(R \pm n_d) \left(\frac{(1-s)}{p} \right) \pm 2n_{sa} \pm n_{ws} \right] f \tag{6.61}$$

with the associated mode (pole pair) number of

$$m = (R \pm S \pm n_s \pm n_d \pm 2n_{sa} p \pm n_\theta p) \tag{6.62}$$

where the n is any integer (including 0) and subscripts $r,d,sa,\omega s,s,\theta$ refer to rotor, dynamic, saturation, time harmonic, stator, and space angle (either stator or rotor), respectively. At the low frequency level, only the increase in mixed eccentricity gives rise to low frequency-related mixed eccentricity components [45,46], such as

$$f_{ecc2} = [f \pm kf_r], k = 1,2,3,...;\ f_r = \frac{(1-s)}{p} f \tag{6.63}$$

where f_r is rotational spped in Hz.

Sometimes these components can increase when one of the eccentricities increase, but to see such an increase the other kind of eccentricity should be high enough. Normally dynamic eccentricity is tightly controlled in a new motor by total indicated reading (TIR), and total eccentricity level is usually kept lower than 10% in order to control the unbalanced magnetic pull (UMP), acoustic noise, and mechanical vibration. Also, secondary effects, such as speed oscillations due to high dynamic eccentricity, can also give rise to these components.

Usually, in literature, a more simplified version of Equation (6.61) (with $n_{sa} = 0$) is quoted. With both $n_d, n_{sa} = 0$ in Equation (6.61) one arrives at the formula of principle slot harmonics (PSH) or rotor slot harmonics (RSH). These components, apart from being used to detect only static eccentricity, can be employed for sensorless speed measurements of induction motors also [44,45]. However, in some of the motors where the number of fundamental poles ($2p$) is an exact divisor of the number of rotor bars according to Equation (6.64),

$$R = 2p[3(m \pm q) \pm r]\ , m \pm q = 0, 1, 2, 3 \ldots, r = 0 \ or \ 1 \tag{6.64}$$

these components may not show sufficient increase for static or dynamic eccentricity. These components will, however, be strong only when a considerable amount of both the eccentricities are present. The relationship between the pole pairs and rotor bars to be a good indicator of static or dynamic eccentricity, is a modified form of Equation (6.64) and is given by

$$R = 2p[3(m \pm q) \pm r] \pm 1 \tag{6.65}$$

This distinction was not made very clear by Cameron et al. [43]. While increase of static eccentricity increased the PSH in the 28 bar, 4 pole machine to an appreciable extent only after it was increased beyond 60%, most of the dynamic-eccentricity-related components increased in this machine almost linearly up to 40% increase in static eccentricity. This led to the conclusion by researchers that dynamic eccentricity could also be a by-product of static eccentricity. This was most probably due to increase in rotor bounce caused by UMP as was shown later by Thomson et al. [47]. Since 28/4 = 7, the number of fundamental poles is an exact divisor of the number of rotor bars, and hence it does not fall in the category defined by Equation (6.65). Therefore the observed increase in the PSH and the dynamic eccentricity components by Cameron et al. [43] can be explained as an increase in *mixed* eccentricity. In fact, the high frequency components similar to Equation (6.61) may not be very strong in machines that do not fall in the category defined by Equation (6.64) and Equation (6.65).

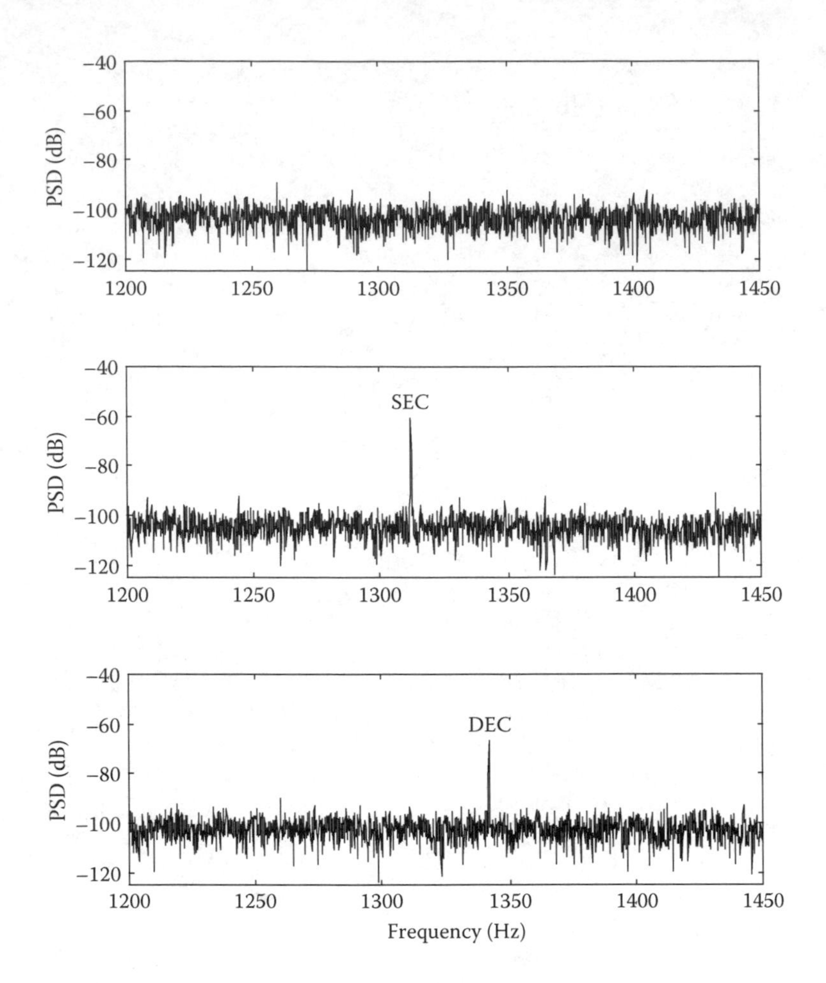

FIGURE 6.25
Simulated, normalized line current spectrum of a 4 pole, 43 bar, 60 Hz induction motor under healthy (top), with 41% static eccentricity (middle) and 20% dynamic eccentricity (bottom). Slip = 0.029. This motor conforms to Equation (6.65).

To illustrate some of the statements made in the previous paragraph, simulation studies were conducted on machines with eccentricity but *without slotting permeance effect* [44,45,48]. However, since each loop of the rotor was individually modeled, the rotor current distribution carried the rotor bar information implicitly (through the rotor current MMF's space harmonics) and therefore could produce the PSH and the other eccentricity-related harmonics quite clearly. Figure 6.25 and Figure 6.26 show these results. Slotting effects were later included in another study and did not show much change in the PSH results under concentric condition of the machine. From this, it was concluded that the rotor MMF effect has considerable effect on the magnitude of these harmonics.

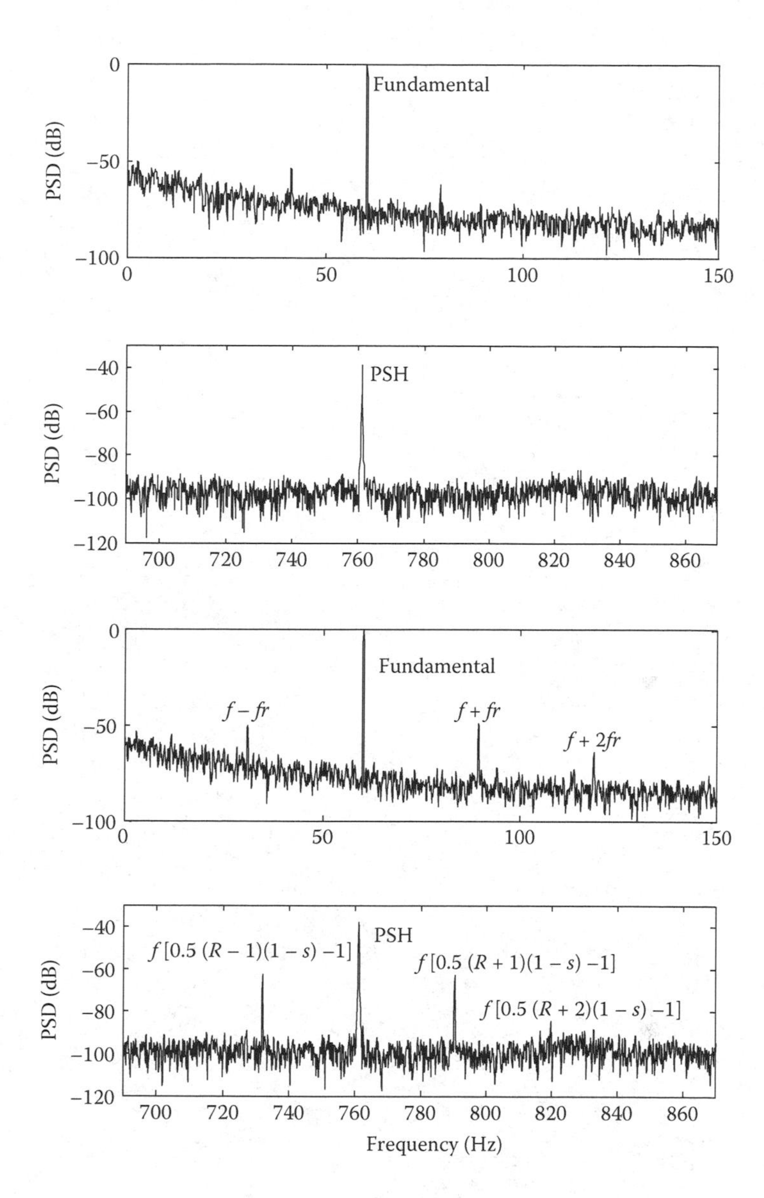

FIGURE 6.26
Simulated, normalized line current spectrum of a 4 pole, 28 bar, 60 Hz induction motor (from top to bottom): healthy around fundamental, healthy around PSH, with 41% static eccentricity and 21% dynamic eccentricity (or mixed eccentricity) around fundamental, with mixed eccentricity around PSH (bottom). Slip = 0.022.

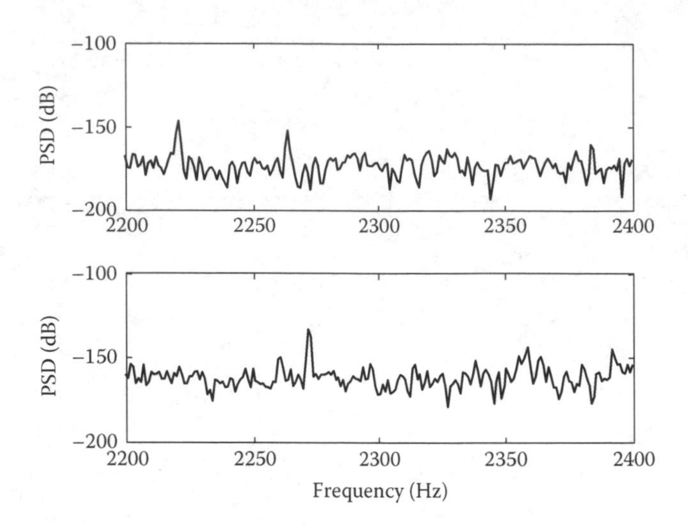

FIGURE 6.27
Experimental, normalized line current spectrum of a 300 hp, 2 pole, 39 bar, 60 Hz induction motor under full load with s = 0.0073 (top) (PSH1 = 2262.93 Hz, PSH2 = 2382.92 Hz) and no load (bottom) s = 0.0013(PSH1 = 2276.96 Hz, PSH2 = 2396.96 Hz).

Experimentally similar effects also showed up [44,45,49]. When a 39 slot, 2 pole machine was tested it showed little PSH (Figure 6.26). So did a 10 pole, 94 bar machine (Figure 6.28). However, a 4 pole, 44 bar machine showed large PSH. Also, its low-frequency as well as high-frequency-related components did not show much change, with 39% static eccentricity and inherent dynamic eccentricity (Figure 6.29). But with around 41% static and 21% dynamic eccentricity introduced in a 4 pole, 28 bar motor, which falls in the same category as the 4 pole, 44 bar machine, most of the predicted low as well as high frequency components showed marked changes as illustrated in Figure 6.30 and Figure 6.31. Curiously, unlike results described by

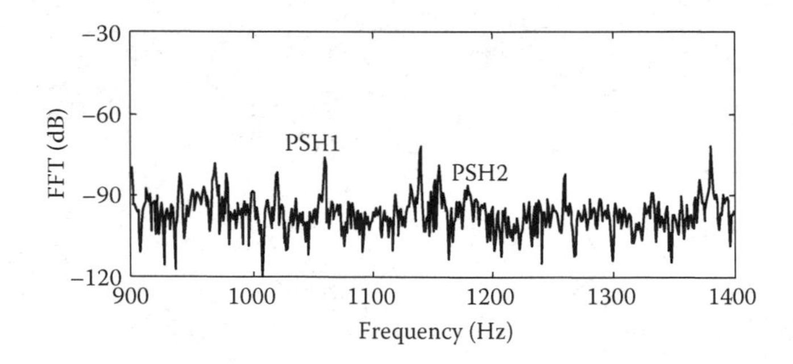

FIGURE 6.28
Experimental, normalized line current spectrum of a 10 pole, 94 bar, 60Hz induction motor under half load with $s = 0.00621$.

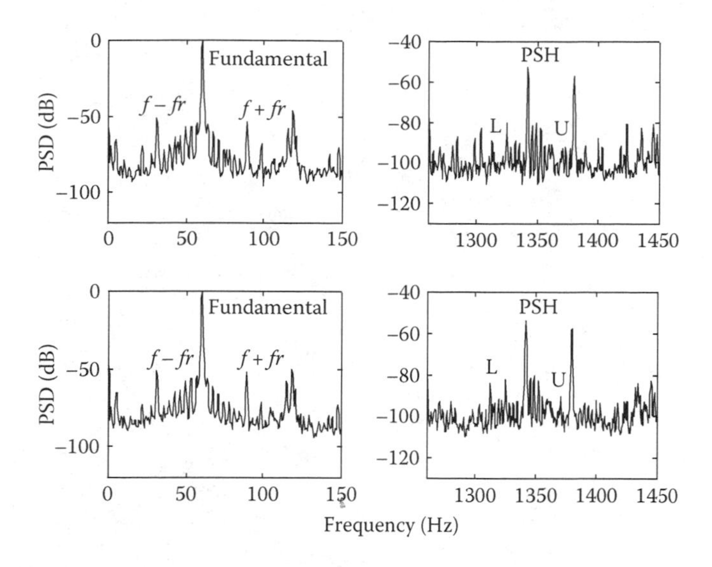

FIGURE 6.29
Experimental, normalized spectra of the line current of machine under load around the fundamental and PSH. Upper row: healthy, lower row: with mixed eccentricity (39% SE, inherent DE). $R = 44$. Slip = 0.029. Lower frequency component and upper frequency component (L&U) show the other dynamic eccentricty components.

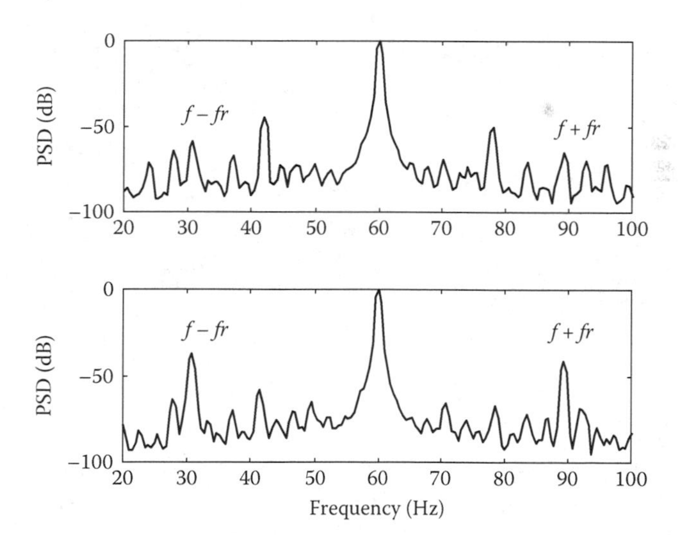

FIGURE 6.30
Experimental, normalized spectra of the line current of machine under load around the fundamental. Upper: healthy, lower: with mixed eccentricity (41% SE, 21% DE). $R = 28$. Slip = 0.022.

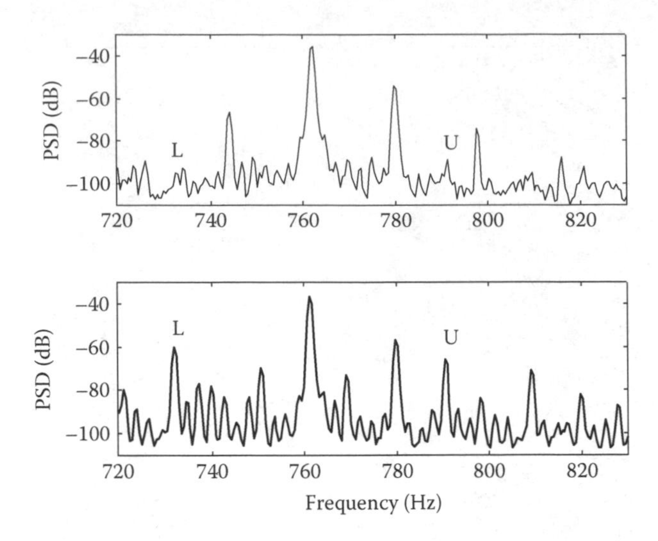

FIGURE 6.31
Experimental, normalized spectra of the line current of machine under full load around the PSH. Upper: healthy, lower: with mixed eccentricity (41% SE, 21% DE). R = 28. Slip = 0.022. Lower frequency component and upper frequency component (L&U) show the other dynamic eccentricuty related components.

Cameron et al. [43], the PSH did not show any noticeable change. This could be because results obtained by Cameron et al. were for a very high value (80%) of static eccentricity. Also, the difference in results could be due to different skewing of slots. Skewing reduces these components to a large degree [50]. However, when these experiments were repeated on a 45 bar, 4 pole, 60 Hz machine, which conform to Equation (6.65), the high-frequency-related static eccentricity components [49] clearly showed up (Figure 6.32).

6.3.4.2 Detection of Eccentricity Faults Based on Nameplate Parameters

It is clear from the discussion in the previous section that it is very difficult to detect eccentricity without the knowledge of rotor slots if high-frequency signals such as Equation (6.61) are used to detect eccentricity faults. Although it is true that with significant eccentricity Equation (6.63) will in general show increase, the real cause of such an increase may not be clear. It was shown by Thomson et al. that static-eccentricity-caused UMP resulted in rotor bounce [47]. The ensuing dynamic eccentricity caused not only an increase in mixed-eccentricity-related signals given by Equation (6.63) but also increased dynamic-eccentricity-related signals given by Equation (6.61). Hence, to isolate the kind of eccentricity, both Equation (6.61) and Equation (6.63) need to be looked at. Unfortunately, rotor slot number is not a nameplate parameter

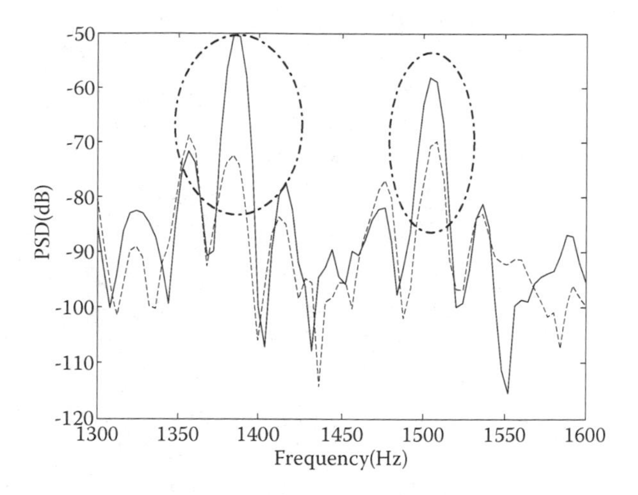

FIGURE 6.32
Comparison of line current spectra of a healthy (dotted) and eccentric machine for the confirmation of the presence of static eccentricity. The spectral peaks circled (from left to right) are 1384Hz (nws = 1) and 1504 (nws = 3) Hz, respectively.

and its value may not be easily obtainable. Also, inherent eccentricity can actually mask eccentricity faults by reducing the magnitudes of components given by Equation (6.63). This is clearly evident from Figure 6.28 where no appreciable change is visible, even with the introduction of substantial static eccentricity. Hence a baseline value of the low frequency sideband is always required. Also pole pair, rotor slot, and skewing play a significant role in obtaining a good spectra given by Equation (6.61).

Recently, a new scheme proposed by Nandi et al. took a fresh look at some of the voltage harmonics at the terminal voltage right after motor switch-off [49]. Analysis showed that only static eccentricity will cause a change in harmonic components given by

$$f_e = \left(\frac{n}{p}\right) pf_r, f_r = \frac{\omega_r}{2\pi}, n = 1, 2, 3... \tag{6.66}$$

whereas only dynamic eccentricity will cause a change in harmonic components given by

$$f_e = \left(\frac{n \pm k_d}{p}\right) pf_r f_r = \frac{\omega_r}{2\pi}, n = 1, 2, 3..., k_d = 1, 2, 3... \tag{6.67}$$

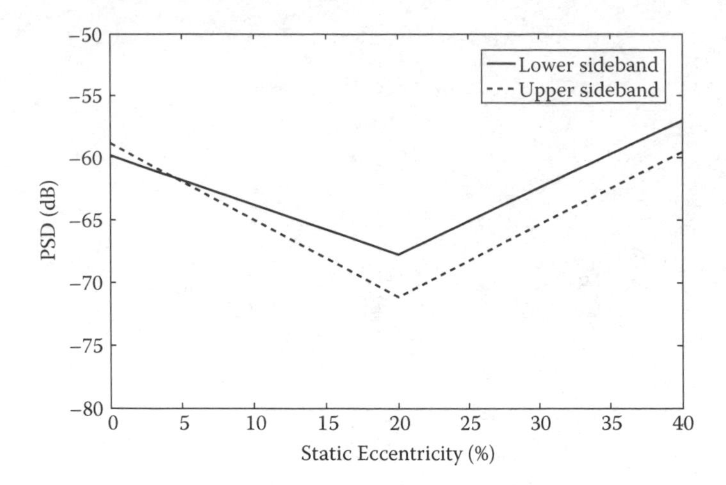

FIGURE 6.33
Change in low frequency sidebands given by (4) with inherent dynamic eccentricity and different levels of static eccentricity in a 4 pole, 44 bar, 7.5 hp motor. 0% static eccentricity implies inherent level of static eccentricity.

In both Equation (6.66) and Equation (6.67), p is the number of pole pairs, ω_r is the speed in radians per second, and k_d is the dynamic eccentricity induced harmonic number. It was observed that static eccentricity preserved the original healthy spectral pattern and caused changes in all the odd harmonics described by Equation (6.66), except for an increased noise floor near the prominent slot harmonics. Due to the reverse rotating field caused by static eccentricity, some odd triplen harmonics also changed more that dynamic eccentricity. Dynamic eccentricity destroyed the original spectral pattern (by flattening it out) near the prominent slot harmonics due to increase in some of the even harmonics, apart from the increase of noise floor around these harmonics. Figure 6.34 and Figure 6.35 clearly show these trends in a 4 pole, 44 bar, 7.5 hp motor. The main advantage of this technique is (1) the detection of ecccentricity faults in individual form, even in machines that do not show these signatures in line current spectrum in steady state, (2) to detect the main contributory factor in case of mixed eccentricity, and (3) complete absence of UMP. Since the stator poles are absent after switch-off, the magnetic pull that exists between the stator and rotor vanishes. This means the static eccentricity induced dynamic eccentricity (due to UMP-caused rotor bounce), as reported by Thomson et al. [47], will be absent. Thus, the problem will be correctly diagnosed as static eccentricity and not as mixed eccentricity using the proposed technique. Even the high frequency spectrum results leading to anomalous identification as reported by Thomson et al. can be corrected with this method. No other existing method can claim this advantage.

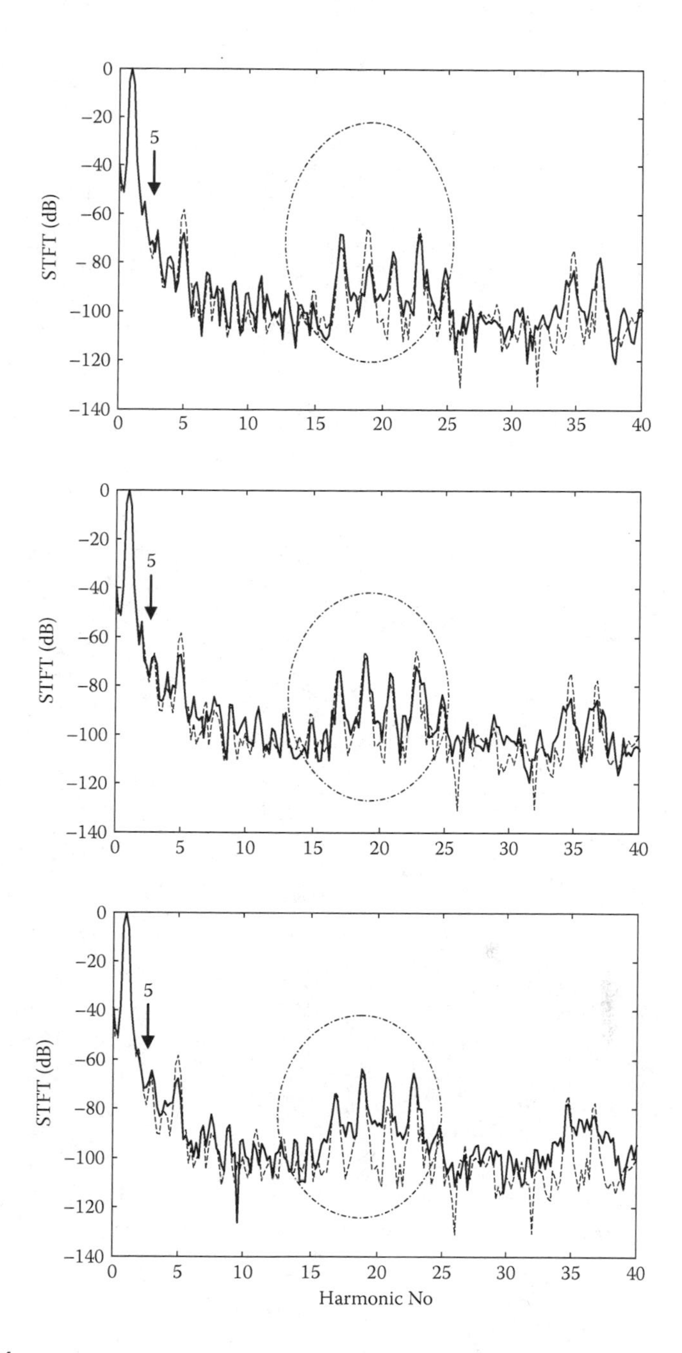

FIGURE 6.34
Static eccentricity related terminal voltage spectrum after switch-off. 20% (top), 40% (middle), and 60% (bottom). Dotted is healthy spectrum. The spectral lines inside the encircled area from left to right correspond to 17th, 19th, 21st, and 23rd harmonics. The spectra are normalized with respect to fundamental (pf_r) at 60.24 Hz (at 0 dB).

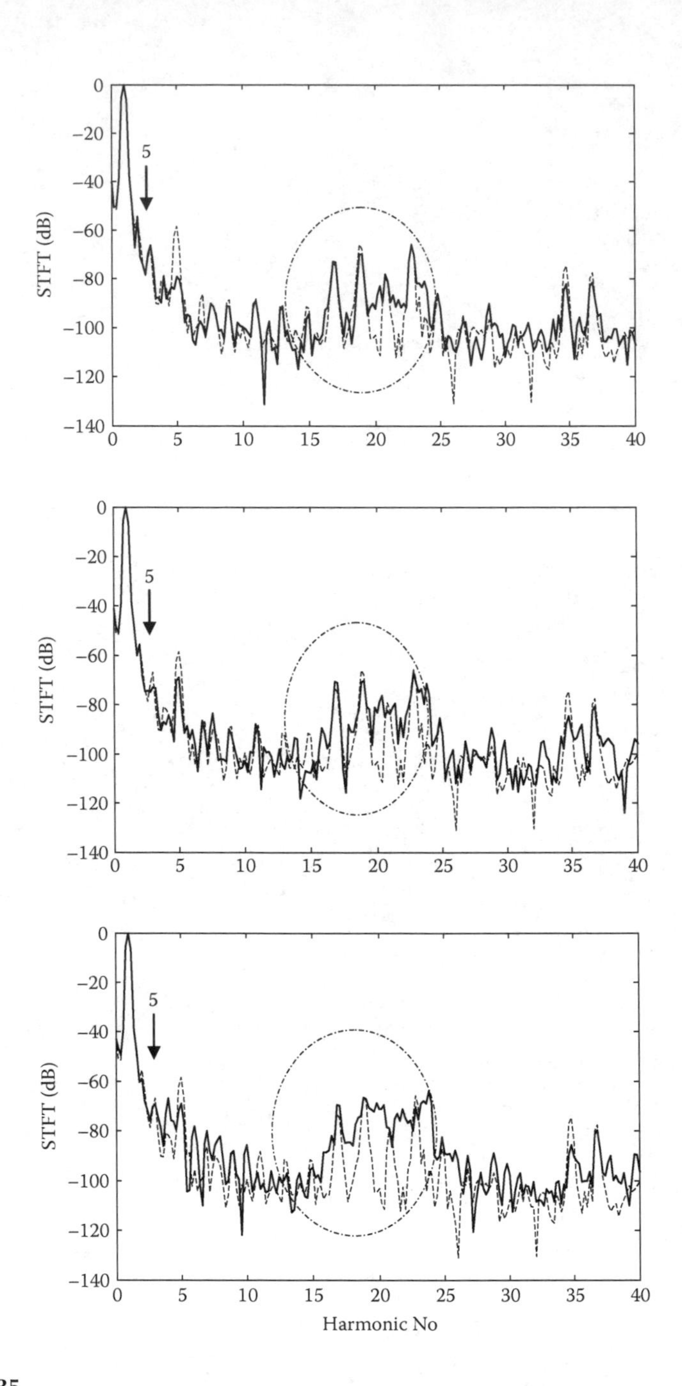

FIGURE 6.35
Dynamic eccentricity related terminal voltage spectrum after switch-off. 20% (top), 40% (middle), and 60% bottom. Dotted is healthy spectrum. The spectral lines inside the encircled area from left to right reflect changes to 20th, 22nd, and 24th harmomics. The spectra are normalized with respect to fundamental (pf_r) at 60.24 Hz (at 0 dB).

6.3.4.3 Detection of Eccentricity Faults Using Mechanical Vibration Signal Spectra

The mechanical vibration frequency components in a spectrum is given by the time-frequency components that appear in the force wave distribution given by

$$\sigma(\theta,t)=\frac{B^2(\theta,t)}{2\mu_0} \tag{6.68}$$

Thus, using a generalized form of $B(\theta,t)$ [43], the following frequencies can increase in the mechanical vibration spectrum due to eccentricity:

$$f_{sv}=\left\{(n'_{rt}R\pm n'_d)\frac{(1-s)}{p}\pm 2n'_{sa}\pm n'_{\omega}\right\}f \tag{6.69}$$

with an associated mode number (similar to pole pair number) of

$$m=n'_{rt}R\pm n'_{st}S\pm n'_s\ \pm n'_d\ \pm 2n'_{sa}p\pm n'_{\theta}\ p \tag{6.70}$$

where the n' is any integer (including 0) and the subscripts rt,d,sa,ω,st,θ refer to rotor, dynamic, saturation, time harmonic, stator and space angle (either stator or rotor), respectively. These frequency components are usually visible for a low mode number. For example, the pole number related to the 886 Hz component is 2.

When both static and dynamic eccentricity is present, using another form of $B(\theta,t)$ not involving rotor slots [46], the increase of the following low frequency components also can be predicted:

$$f_{lv}=2f\pm f_r \tag{6.71}$$

where f and f_r are the supply and rotational frequencies. For a certain dynamic eccentricity, the $2f+f_r$ component increases more sharply with static eccentricity than the $2f-f_r$ component. The $2f$ component is also expected to rise with the increase of only static eccentricity.

6.3.4.4 Detection of Inclined Eccentricity Faults

So far our discussion was limited to a uniform level of eccentricity. However, it is quite possible that the level of eccentricity may not be uniform. For example, the load-side bearing in a motor may encounter more wear and tear than the drive-end bearing, leading to nonuniform eccentricity. The effect of inclined eccentricity was studied extensively on a 4 pole,

TABLE 6.3

Experimental, Normalized Amplitude of Eccentricity-Related Harmonics Arising out of Pole Pair Matching

	Amplitude of Eccentricity-Related Harmonics for Inclined Condition			
Load Level	**50%, 50%**	**45.78%, 65.06%**	**Healthy**	**50%, –50%**
0%	–49.26 dB	–50.74 dB	–61.01 dB	–62.7 dB
25%	–48.97 dB	–49.55 dB	–64.6 dB	–61.08dB
50%	–47.22 dB	–48.27 dB	–63.8 dB	–61.24 dB
75%	–46.9 dB	–47.38 dB	–62.42 dB	–61.7 dB
100%	–46.88 dB	–47.28 dB	–62.04 dB	–61.9 dB

45 bar induction motor [51]. This motor conforms to Equation (6.65). It was observed that the increase in the spectral components due to inclined eccentricity is similar to that which would be produced by a uniform eccentricity equal to the average value of the inclined eccentricity. Also only one of the PSH arising out of the pole pair matching showed increase. Most important, if the eccentricity at one end is equal and opposite to the eccentricity at the other end, then the eccentricity may not be detectable at all, even though a stator–rotor rub may be imminent. Subsequently, even the mechanical vibration signals remained inconclusive in detecting the equal but opposite eccentricity. Table 6.3 and Table 6.4 list the current spectra results under different load conditions.

6.3.5 Detection of Faults in Inverter-Fed Induction Machines

Low-order harmonic components that appear in the line voltage of inverters supplying induction motors can provide additional information on the fault of induction machines. This was illustrated very clearly by Akin et al. [52] for bearing, eccentricity, and broken rotor bar faults. The fault frequencies to

TABLE 6.4

Experimental, Normalized Amplitude of Eccentricity-Related Harmonics Arising out of Asymmetry

	Amplitude of Asymmetry-Related Harmonics for Inclined Condition			
Load Level	**50%, 50%**	**45.78%, 65.06%**	**Healthy**	**50%, –50%**
0%	–51.3 dB	–52.2 dB	–55.94 dB	–57.45 dB
25%	–53.7 dB	–55.4 dB	–56.54 dB	–56.6 dB
50%	–55.2 dB	–58.9 dB	–59.4 dB	–58.55 dB
75%	–55.9 dB	–57.38 dB	–59.6 dB	–59.2 dB
100%	–55.8 dB	–57.28 dB	–60.2 dB	–59.25 dB

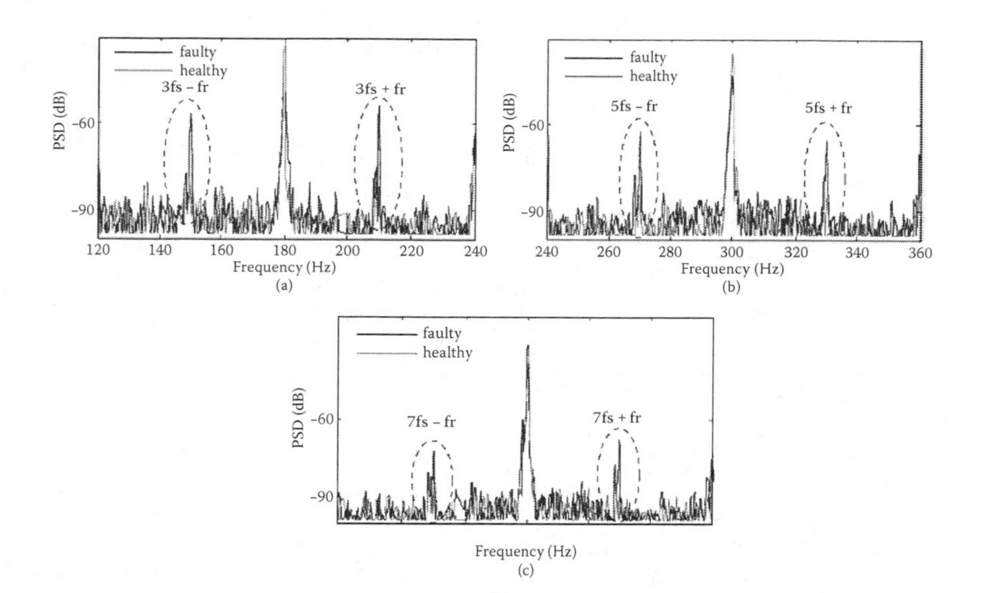

FIGURE 6.36
Current spectrum of the inverter fed healthy and eccentric 4 pole induction motor (for f_s = 60 Hz).

be detected can be simply derived by replacing the fundamental frequency with these harmonics in the equations describing these frequencies. For example, if Equation (6.63) is to be used for these low frequency spectra for a harmonic of order h, then it is to rewritten as

$$f_{ecc2} = [hf_s \pm kf_r], h = 1,3,5,7...k = 1,2,3,... \tag{6.72}$$

where f_s is the fundamental supply harmonic. Figure 6.36 illustrates the concept.

References

[1] J.H. McClellan, R.W. Schafer, and M.A. Yoder, *Signal Processing First*, New York: Pearson Prentice Hall, 2003.
[2] R.A. Roberts and C.T. Mullis, *Digital Signal Processing*, Reading, MA: Addison-Wesley, 1987.
[3] J.G. Proakis and D.G. Manolakis, *Digital Signal Processing: Principles, Algorithms, and Applications*, New Delhi: Prentice Hall, 1997.

[4] A. Yazidi, H . Henao, G.A. Capolino, M. Artioli, and F. Filippetti, "Improvement of frequency resolution for three-phase induction machine fault diagnosis," *Industrial Application Annual Meeting Conference*, pp. 21–25, Hong Kong, China, October 2–6, 2005.
[5] M.E. Benbouzid, M. Vieira, and C. Theys, "Induction motors' fault detection and localization using stator current advanced signal processing techniques," *IEEE Transactions on Power Electronics*, vol. 14, no. 1, pp. 14–22, January 1999.
[6] M.E. Benbouzid and G.B. Kliman, "What stator current processing-based technique to use for induction motor faults diagnosis?" *IEEE Transactions on Energy Conversion*, vol. 18, no. 2, pp. 238–244, June 2003.
[7] MATLAB 7.2 R2006a, Natick, MA: The Mathworks Inc., 2006.
[8] N. Arthur and J. Penman, "Induction machine condition monitoring with higher order spectra," *IEEE Transactions on Industrial Electronics*, vol. 47, no. 5, pp. 1031–1041, October 2000.
[9] X.M. Gao, S.J. Ovaska, A. Shenghe, and Y.C. Jenq, "Analysis of second-order harmonic distortion of ADC using bispectrum," *IEEE Transactions on Instrumentation and Measurement*, vol. 45, no. 1, pp. 50–55, February 1996.
[10] Stanford Research Systems, *Operating manual and programming Reference SR785—Dynamic Signal Analyzer Users Manual*, Sunnyvale, CA, September 2004.
[11] E.J. Wiedenburg, "Measurement Analysis and Efficiency Estimation of Three Phase Induction Machines Using Instantaneous Electrical Quantities," PhD dissertation, Oregon State University, 1998.
[12] R.R. Obaid, T.G. Habetler, and J.R. Stack, "Stator current analysis for bearing damage detection in induction motors," *SDEMPED 2003*, pp. 182–187, August 21–26, 2003.
[13] J.S. Stack, T.G. Habetler, and R.G. Harley, "Fault classification and fault signature production for rolling element bearings in electric machines," *IEEE Transactions on Industrial Application*, vol. 40, no. 3, pp. 735–739, May/June 2004.
[14] S.A. McInerny and Y. Dai, "Basic vibration signal processing for bearing fault detection," *IEEE Transactions on Education*, vol. 46, no. 1, pp. 149–156, February 2003.
[15] B. Li, M.-Y. Chow, Y. Tipusuwan, and James C. Hung, "Neural-network-based motor rolling bearing fault diagnosis," *IEEE Transactions on Industrial Electronics*, vol. 47, no. 5, pp. 1060–1069, October 2000.
[16] J.S. Stack, T.G. Habetler, and R.G. Harley, "An amplitude modulation detector for fault diagnosis in rolling element bearings," *IEEE Transactions on Industrial Electronics,* vol. 51, no. 5, pp. 1097–1102, October 2004.
[17] J.S. Stack, T.G. Habetler, and R.G. Harley, "Fault-signature modeling and detection of inner-race bearing faults," *IEEE Transactions on Industrial Applications*, vol. 42, no. 1, pp. 735–739, January/February 2006.
[18] J.S. Stack, T.G. Habetler, and R.G. Harley, "Experimentally generating faults in rolling element bearings via shaft current," *IEEE Transactions on Industrial Applications*, vol. 41, no.1, pp. 25–29, January/February 2005.
[19] R.R. Schoen, T.G. Habetler, F. Kamran, and R.G. Barthold, "Motor current damage detection using stator current monitoring," *IEEE Transactions on Industrial Applications*, vol. 31, no. 6, pp. 25–29, November/December 1995.

[20] J.S. Stack, T.G. Habetler, and R.G. Harley, "Bearing fault detection via autoregressive stator current modeling," *IEEE Transactions on Industrial Applications*, vol. 40, no. 3, pp. 25–29, May/June, 2004.
[21] J. Penman, M.N. Dey, A.J. Tait, and W.E. Bryan, "Condition monitoring of electrical drives," *Proceedings of IEEE*, vol. 133, pt. B, pp. 142–148, May 1986.
[22] J. Penman, H.G. Sedding, B.A. Lloyd, and W.T. Fink, "Detection and location of interturn short circuits in the stator windings of operating motors," *IEEE Transactions on Energy Conversion*, vol. 9, no. 4, pp. 652–658, December 1994.
[23] H. Henao, C. Demian, and G.A. Capolino, "A frequency-domain detection of stator winding faults in induction machines using an external flux sensor," *IEEE Transactions on Industrial Applications*, vol. 39, no. 5, pp. 1272–1279, September/October 2003.
[24] A. Stavrou, H. Sedding, and J. Penman, "Current monitoring for detecting interturn short circuits in induction motors," *IEEE Transactions on Energy Conversion*, vol. 16, no. 1, pp. 32–37, March 2001.
[25] G. Joksimovic and J. Penman, "The detection of interturn short circuits in the stator windings of operating motors," *IEEE Transactions on Industrial. Electronics*, vol. 47, no. 5, pp. 1078–1084, October 2000.
[26] S.M.A. Cruz and A.J.M. Cardoso, "Diagnosis of stator inter-turn short circuits in DTC induction motor drives," *IEEE Transactions on Industrial Application*, vol. 40, no. 5, pp. 1349–1360.
[27] S. Nandi, "A detailed model of induction machines with saturation extendable for fault analysis," *IEEE Transactions on Industrial Application*, vol. 40, no. 5, pp. 1302–1309, September/October 2004.
[28] S. Nandi and H.A. Toliyat, "A novel frequency domain-based technique to detect stator inter-turn faults in induction machines using stator induced voltages after switch-off," *IEEE Transactions on Industrial Application*, vol. 38, no. 1, pp. 101–109, January/February 2002.
[29] S. Nandi, "Detection of stator faults in induction machines using residual saturation harmonics," *IEEE Transactions on Industry Applications*, vol. 42, no. 5, pp. 1201–1205, September/October 2006.
[30] P. Neti and S. Nandi, "Detection of stator inter-turn faults in reluctance synchronous motor by using stored magnetic energy after supply disconnection," *Proceedings of the National Power Electronics Conference*, pp. 45–48, Kharagpur, India, December 22–24, 2005.
[31] P. Neti and S. Nandi, "Stator inter-turn fault detection of synchronous machines using field current and rotor search coil voltage signature analysis," *IEEE Transactions on Industry Applications*, vol. 45, no. 3, pp. 911–920, May/June 2009.
[32] D. Shah, S. Nandi, and P. Neti, "Stator inter-turn fault detection of doubly-fed induction generators using rotor current and search coil voltage analysis," *IEEE Transactions on Industry Applications*, vol. 45, no. 5, pp. 1831–1842, September/October 2009.
[33] R.F. Walliser and C.F. Landy, "Determination of interbar current effects in the detection of broken rotor bars in squirrel cage induction motors," *IEEE Transactions on Energy Conversion*, vol. 9, no. 1, pp. 152–158, March 1994.

[34] G.B. Kliman, R.A. Koegl, J. Stein, R.D. Endicott, and M.W. Madden, "Noninvasive detection of broken rotor bars in operating induction motors," *IEEE Transactions on Energy Conversion*, vol. 3, no. 4, pp. 873–879, December 1988.
[35] S. Nandi, "Fault Analysis for Condition Monitoring of Induction Motors," PhD dissertation, Texas A&M University, May 2000.
[36] F. Filippetti, G. Franceschini, C. Tassoni, and P. Vas, "AI techniques in induction machines diagnosis including the speed ripple effect," *IEEE Transactions on Industrial Applications*, vol. 34, no. 1, pp. 98–108, January/February 1998.
[37] G. Didier, E. Ternisien, O. Caspary, H. Razik, H. Henao, A. Yazidi, and G.-A. Capolino, "Rotor fault detection using the instantaneous power signature industrial technology," *IEEE ICIT '04*, vol. 1, pp. 170–174, December 2004.
[38] N.M. Elkasabgy, A.R. Eastham, and G.E. Dawson, "Detection of broken bars in the cage rotor on an induction machine," *IEEE Transactions on Industry Applications*, vol. 22, no. 6, pp. 165–171, January/February 1992.
[39] H. Meshgin-Kelk, J. Milimonfared, and H.A. Toliyat, "Interbar currents and axial fluxes in healthy and faulty induction motors," *IEEE Transactions on Industrial Application*, vol. 40, no. 1, pp. 128–134, January/February 2004.
[40] J. Milimonfared, H. Meshgin-Kelk, S. Nandi, A. Der Minassians, and H.A. Toliyat, "A novel approach for broken rotor bar detection in cage induction motors," *IEEE Transactions on Industrial Applications*, vol. 35, no. 5, pp. 1000–1006, September/October 1999.
[41] B. Akin, S.B. Ozturk, H.A. Toliyat, and M. Rayner, "DSP-based sensorless electric motor fault-diagnosis tools for electric and hybrid electric vehicle power train applications," *IEEE Transactions of Vehicular Technology*, vol. 58, no. 5, pp. 2150–2159, July 2009.
[42] G.H. Muller and C.F. Landy, "A novel method to detect broken bars in squirrel cage induction motors when interbar currents are present," *IEEE Transactions on Energy Conversion*, vol. 18, no. 1, pp. 71–79, March 2003.
[43] J.R. Cameron, W.T. Thomson, and A.B. Dow, "Vibration and current monitoring for detecting air gap eccentricity in large induction motors," *IEEE Proceedings*, vol. 133, pt. B, no. 3, pp. 155–163, May 1986.
[44] S. Nandi, S. Ahmed, H.A. Toliyat, and R.M. Bharadwaj, "Selection criteria of induction machines for speed-sensorless drive applications," *IEEE Transactions on Industry Applications*, vol. 39, no. 3, pp. 704–712, May/June 2003.
[45] S. Nandi, R. Bharadwaj, and H.A. Toliyat, "Performance analysis of a three-phase induction motor under incipient mixed eccentricity condition," *IEEE Transactions on Energy Conversion*, vol. 17, no. 3, pp. 392–399, September 2002.
[46] D.G. Dorrell, W.T. Thomson, and S. Roach, "Analysis of air-gap flux, current, vibration signals as a function of the combination of static and dynamic air-gap eccentricity in 3-phase induction motors," *IEEE Transactions on Industry Applications*, vol. 33, no. 1, pp. 24–34, 1997.
[47] W.T. Thomson, D. Rankin, and D.G. Dorrell, "On-line current monitoring to diagnose airgap eccentricity in large three-phase induction motors—Industrial case histories verify the predictions," *IEEE Transactions on Energy Conversion*, vol. 14, no. 4, pp.1372–1378, December 1999.
[48] S. Nandi, X. Li, and T. Ilamparithi, "Recent developments in the modeling and analysis of induction machines with non-uniform air-gap using the modified winding function approach," *ICAECT Conference*, pp. 182–187, January 2010.

[49] S. Nandi, T. Ilamparithi, S.B. Lee, and D. Hyun, "Detection of eccentricity faults in induction machines based on nameplate parameters," *IEEE Transactions on Industrial Electronics,* vol. 58, no. 5, pp. 1673–1683, May 2011.

[50] W.T. Thomson and A. Barbour, "On-line current monitoring and application of a finite element method to predict the level of static airgap eccentricity in three-phase induction motors," *IEEE Transactions on Energy Conversion*, vol. 13, no. 4, pp. 347–357, December 1998.

[51] X. Li and S. Nandi, "Performance analysis of a 3-phase induction machine with inclined static eccentricity," *IEEE IEMDC Conference*, pp. 1606–1613, May 2005.

[52] B. Akin, U. Orguner, H.A. Toliyat, and M. Rayner, "Low order PWM inverter harmonics contributions to the inverter-fed induction machine fault diagnosis," *IEEE Transactions on Industrial Electronics*, vol. 55, no. 2, pp. 610–619, February 2008.

7

Fault Diagnosis of Electric Machines Using Model-Based Techniques

Subhasis Nandi, Ph.D.
University of Victoria

7.1 Introduction

Fault detection implies a two-valued outcome depending upon the normal or abnormal operating characteristics of a system [1]. Fault diagnosis is the process that actually decides the cause, nature, and location of a fault. Incipient fault diagnosis may even be the preemptive process to minimize damage due to faults. To make a fault diagnosis scheme incipient, it requires monitoring the system at every instant. The most logical way to implement this is to compare the system outputs with set reference values. This could be based on three possible ways: (1) signal, (2) knowledge, and (3) model. In a signal-based approach, the outputs are compared with average or limit values. It is very simple to apply. However, its use for early detection or trend monitoring is very limited.

Knowledge-based methods usually depend upon qualitative process structure, functions, and qualitative models to predict fault. Model-based techniques use analytical models of the process to generate "normal outputs" that are compared with the actual process outputs to generate "residuals" that are ultimately used for fault detection. A very simple model-based fault detection scheme is shown in Figure 7.1 [2]. The analytical models can be mathematical models, or generic models using neural networks as shown in Figure 7.2 [1], fuzzy logic presented in Figure 7.3 [3], or genetic algorithm. These generic models are then trained with healthy and faulty data obtained from real systems. Once trained, they can generate the residuals reliably to detect faults. As an example, the fault diagnosis and detection scheme for induction motor faults shown in Figure 7.2 uses three-phase line voltages $V^{NS}(t)$, currents $I^{NS}(t)$, and speed $\omega^{NS}(t)$ that are essentially nonstationary. Using the present values of voltage, speed, and past current predictions, current predicted values of current $\hat{I}^{NS}(t)$ are generated using a multistep ahead neural network predictor that has been trained to emulate a healthy motor. The residuals $r^{NS}(t)$ are then formed by comparing the actual values with

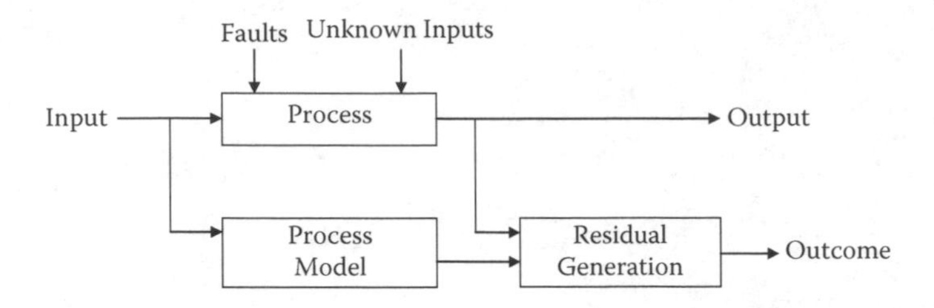

FIGURE 7.1
A general model-based fault detection scheme. (From F. Fischer et al., "Explicit modeling of the stator winding bar water cooling for model-based fault diagnosis of turbogenerators with experimental verification," *Proceedings of the 3rd IEEE Conference on Control Applications*, pp. 1403–1408, August 1994. With permission.)

the predicted values. The residuals $r^{NS}(t)$ are further processed along with the currents $I^{NS}(t)$ to separate the currents and residuals into their fundamental $(I_f^{NS}(t), r_f^{NS}(t))$ and harmonic $(I_h^{NS}(t), r_h^{NS}(t))$ components using a wavelet decomposition algorithm. These components are then used to generate two decoupled indicators: (1) $S(\cdot)$, the root mean square value of the normalized harmonics of the residual to detect mechanical faults and (2) $r^-(\cdot)$, the negative sequence component of the residual to detect electrical faults. This also provides a broad classification of fault category.

Model-based fault diagnosis techniques are finding increasing importance for condition-based maintenance (CBM) rather than scheduled or preventive maintenance. CBM is perceived as the preferred technique when scheduled maintenance or routine machine replacement is not required. Even for systems where scheduled maintenance is desirable, early detection using model-based techniques provides the flexibility to stop operation anytime for preventing catastrophic failures and subsequent damages, fatalities, economic, and legal fallout.

In this chapter we will describe simple linear circuit theory based mathematical models used to predict electrical machine faults. Other types of models using finite element (FE) magnetic circuit equivalents and artificial intelligence (AI) have been already described in other related chapters. Although the models may not always be strictly used the way model-based fault diagnosis systems are designed, the insight and information available from studying these models can be immense. The inferences derived from them have been extensively used to fine-tune signal-based, knowledge-based, and other types of model-based fault diagnosis techniques. The accuracy of these models are usually not very good; hence the users should be well aware of their limitations and to what extent they are being used.

Before we deal with electric motors with fault, it is essential that we deal with healthy motors. We will start with a discussion about the healthy induction motor model and then describe how the different types of faults can be implemented in them. We will also discuss synchronous machine fault models later in the chapter.

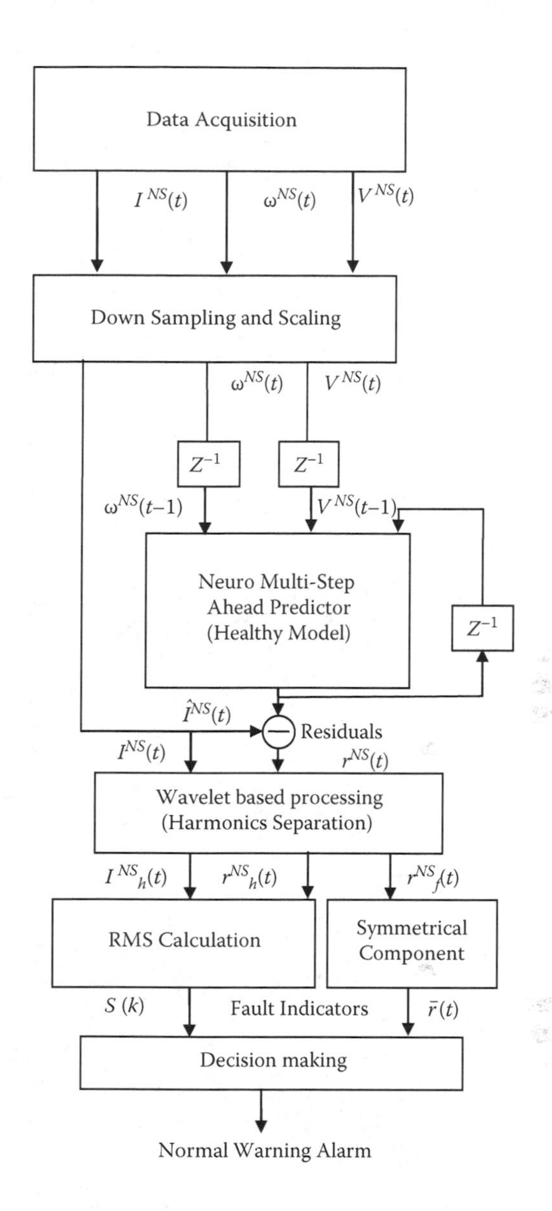

FIGURE 7.2
Neural-network-model-based electric motor fault detection. (From K. Kim and A. Parlos, "Induction motor fault diagnosis based on neuropredictors and wavelet signal processing," *IEEE/ASME Transactions on Mechatronics*, vol. 7, no. 2, pp. 201–219, June 2002. With permission.)

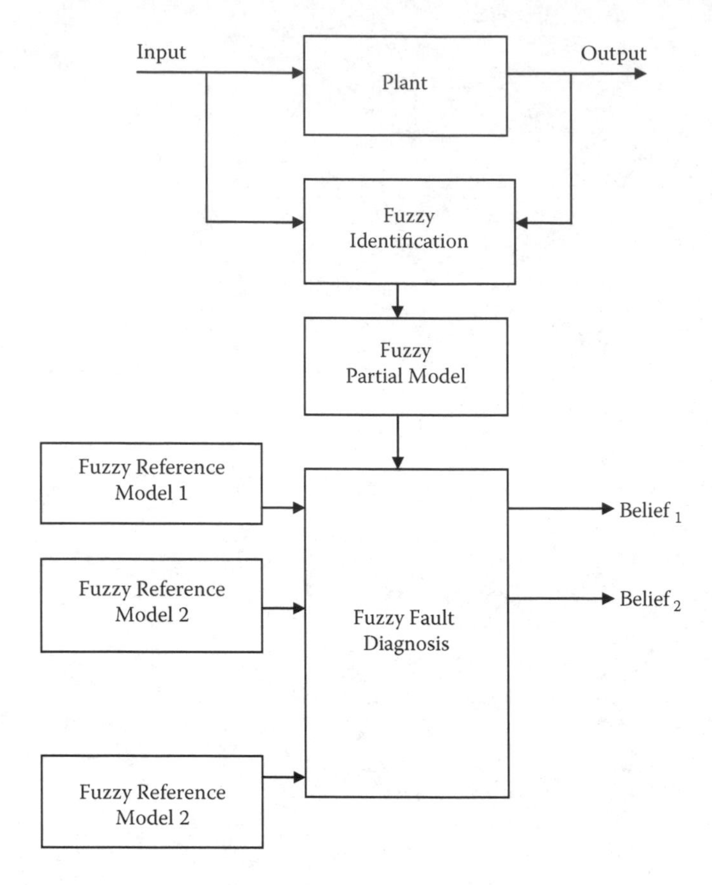

FIGURE 7.3
Fuzzy-logic-based fault diagnosis. (From K. Kim and A. Parlos, "Induction motor fault diagnosis based on neuropredictors and wavelet signal processing," *IEEE/ASME Transactions on Mechatronics*, vol. 7, no. 2, pp. 201–219, June 2002. With permission.)

7.2 Model of Healthy Three-Phase Squirrel-Cage Induction Motor

A squirrel-cage induction motor consists of a stator with a symmetrical multiphase winding and a squirrel-cage with many bars placed at equal distance from one another and shorted at the two ends by two circular slip rings [4]. For the present case we will consider a three-phase, star-connected, single-circuit stator winding with n rotor bars. The model also makes the following assumptions:

1. The motor is unsaturated.
2. It has negligible eddy current, hysteresis, friction, and windage losses.
3. It has insulated rotor bars.

Following simple circuit theory principles, with the star junction voltage as v_s, the instantaneous phase voltage and current relationship for the three-stator phases can then be written as

$$v_a = R_{as} i_{as} + \frac{d\lambda_{as}}{dt} + v_s \tag{7.1}$$

$$v_b = R_{bs} i_{bs} + \frac{d\lambda_{bs}}{dt} + v_s \tag{7.2}$$

$$v_c = R_{cs} i_{cs} + \frac{d\lambda_{cs}}{dt} + v_s \tag{7.3}$$

where the flux linkages are given by

$$\lambda_{as} = L_{as}\, \bar{i_s} + L_{ar}\, \bar{i_r} \tag{7.4}$$

$$\lambda_{bs} = L_{bs}\, \bar{i_s} + L_{br}\, \bar{i_r} \tag{7.5}$$

$$\lambda_{cs} = L_{cs}\, \bar{i_s} + L_{cr}\, \bar{i_r} \tag{7.6}$$

and inductance matrices are defined by

$$L_{as} = [L_{aa}\, L_{ab}\, L_{ac}] \tag{7.7}$$

$$L_{bs} = [L_{ba}\, L_{bb}\, L_{bc}] \tag{7.8}$$

$$L_{cs} = [L_{ca}\, L_{cb}\, L_{cc}] \tag{7.9}$$

$$L_{ar} = [L_{ar_1}\, L_{ar_2} \cdots L_{ar_{n+1}}] \tag{7.10}$$

$$L_{br} = [L_{br_1}\, L_{br_2} \ldots . L_{br_{n+1}}] \tag{7.11}$$

$$L_{cr} = [L_{cr_1}\, L_{cr_2} \ldots . L_{cr_{n+1}}] \tag{7.12}$$

and the stator and rotor currents vectors are given by

$$\bar{i_s} = [i_{as}\, i_{bs}\, i_{cs}]' \tag{7.13}$$

$$\bar{i_r} = [i_{r_1}\, i_{r_2} \cdots i_{r_{n+1}}]' \tag{7.14}$$

The subscript $n + 1$ in Equations (7.10) to (7.14) refers to the end-ring. Since there are n rotor bars, $2n$ end-ring segments and $2n$ nodes, $n + 2n - 2n + 1$ or $n + 1$ independent voltage loop equations can be written. Also, since in a star-connected machine

$$i_{as} + i_{bs} + i_{cs} = 0 \tag{7.15}$$

we can eliminate i_{cs} from Equations (7.1) to (7.3) and write

$$\begin{aligned} v_{ab} = v_a - v_b = R_{as} i_{as} - R_{bs}\, i_{bs})i_{bs} + (L_{aa} - L_{ac} - L_{ba} + L_{bc})\frac{di_{as}}{dt} \\ + (L_{ab} - L_{ac} - L_{bb} + L_{bc})\frac{di_{bs}}{dt} + \omega\frac{\partial(L_{ar} L_{br})}{\partial\theta}\bar{i}_r + (L_{ar} - L_{br})\frac{d\bar{i}_r}{dt} \end{aligned} \tag{7.16}$$

and

$$\begin{aligned} v_{bc} = v_b - v_c = R_{cs} i_{as} + (R_{bs} + R_{cs})i_{bs} + (L_{ba} - L_{bc} - L_{ca} + L_{cc})\frac{di_{as}}{dt} \\ + (L_{bb} - L_{bc} - L_{cb} + L_{cc})\frac{di_{bs}}{dt} + \omega\frac{\partial(L_{br} - L_{cr})}{\partial\theta}\bar{i}_r + (L_{br} - L_{cr})\frac{d\bar{i}_r}{dt} \end{aligned} \tag{7.17}$$

where θ and ω are the angular position and the speed of the rotor, respectively. This way one state variable and v_s can be eliminated in the final solution. Similarly, for the rotor loops (comprising two bars and two portions of the end-rings) we can write

$$\bar{v}_r = R_r\,\bar{i}_r + \frac{d\lambda_r}{dt} \tag{7.18}$$

where

$$\lambda_r = L_r\,\bar{i}_r + L_{rs}\,\bar{i}_s \tag{7.19}$$

and

$$R_r = \begin{bmatrix} 2(R_b + R_e) & -R_b & 0 & \cdots & 0 & -R_b & -R_e \\ -R_b & 2(R_b + R_e) & -R_b & \cdots. & 0 & 0 & -R_e \\ \vdots & \vdots & \vdots & \cdots & \vdots & \vdots & \vdots \\ \vdots & \vdots & \vdots & \cdots & \vdots & \vdots & \vdots \\ 0 & 0 & 0 & \cdots & 2(R_b + R_e) & -R_b & -R_e \\ -R_b & 0 & 0 & \cdots & -R_b & 2(R_b + R_e) & -R_e \\ -R_e & -R_e & -R_e & \cdots & -R_e & -R_e & nR_e \end{bmatrix} \tag{7.20}$$

$$L_r = \begin{bmatrix} L_{mr}+2(L_b+L_e) & L_{r_1r_2}-L_b & L_{r_1r_3} & \cdots & L_{r_1r_{n-1}} & L_{r_1r_n}-L_b & -L_e \\ L_{r_2r_1}-L_b & L_{mr}+2(L_b+L_e) & L_{r_2r_3}-L_b & \cdots & L_{r_2r_{n-1}} & L_{r_2r_n} & -L_e \\ \vdots & \vdots & \vdots & \cdots & \vdots & \vdots & \vdots \\ \vdots & \vdots & \vdots & \cdots & \vdots & \vdots & \vdots \\ L_{r_{n-1}r_1} & L_{r_{n-1}r_2} & L_{r_{n-1}r_3} & \cdots & L_{mr}+2(L_b+L_e) & L_{r_{n-1}r_n}-L_b & -L_e \\ L_{r_nr_1}-L_b & L_{r_nr_2} & L_{r_nr_3} & \cdots & L_{r_nr_{n-1}}-L_b & L_{mr}+2(L_b+L_e) & -L_e \\ -L_e & -L_e & -L_e & \cdots & -L_e & -L_e & nL_e \end{bmatrix} \quad (7.21)$$

$$L_{rs} = [L_{ra}\ L_{rb}\ L_{rc}] \quad (7.22)$$

$$L_{ra} = \begin{bmatrix} L_{r_1a} \\ L_{r_2a} \\ \vdots \\ L_{r_{n+1}a} \end{bmatrix} \quad (7.23)$$

$$L_{rb} = \begin{bmatrix} L_{r_1b} \\ L_{r_2b} \\ \vdots \\ L_{r_{n+1}b} \end{bmatrix} \quad (7.24)$$

$$L_{rc} = \begin{bmatrix} L_{r_1c} \\ L_{r_2c} \\ \vdots \\ L_{r_{n+1}c} \end{bmatrix} \quad (7.25)$$

Using Equation (7.15), i_{cs} can be eliminated from Equation (7.18) and the resulting equation can be written as

$$\bar{v}_r = R_r\,\bar{i}_r + (L_{ra}-L_{rb})\frac{di_{as}}{dt} + (L_{rb}-L_{rc})\frac{di_{bs}}{dt} + \omega\frac{\partial(L_{ra}-L_{rc})}{\partial\theta}i_{as}$$
$$+\,\omega\frac{\partial(L_{rb}-L_{rc})}{\partial\theta}i_{bs} + L_r\frac{di_r}{dt} \quad (7.26)$$

The electromechanical equation can be written as

$$J\frac{d\omega}{dt} = T_m - T_l \quad (7.27)$$

where

$$T_m = 0.5\bar{i}_s^t \frac{\partial L_{ss}}{\partial \theta}\bar{i}_s + 0.5\bar{i}_s^t \frac{\partial L_{sr}}{\partial \theta}\bar{i}_r + 0.5\bar{i}_r^t \frac{\partial L_{rs}}{\partial \theta}\bar{i}_s + 0.5\bar{i}_r^t \frac{\partial L_{rr}}{\partial \theta}\bar{i}_r \tag{7.28}$$

and

$$L_{sr} = \begin{bmatrix} L_{ar} \\ L_{br} \\ L_{cr} \end{bmatrix} \tag{7.29}$$

$$\frac{d\theta}{dt} = \omega \tag{7.30}$$

Equations (7.16), (7.17), (7.26), and (7.28) then can be combined in the state-space form as

$$\dot{x} = Ax + Bu \tag{7.31}$$

where

$$x = \left[i_{as} \; i_{bs} \; \bar{i}_r \; \omega \; \theta \right]^t \tag{7.32}$$

$$u = \left[v_{ab} \; v_{bc} \; 0 \cdots 0 \; \frac{(T_m - T_l)}{J} \; 0 \right]^t \tag{7.33}$$

Equation (7.33) assumes that $\bar{v}_r$ is a null vector since all the rotor loops are short-circuited.

$$A = -A_1^{-1} A_2 \tag{7.34}$$

$$B = A_1^{-1} \tag{7.35}$$

$$A_1 = \begin{bmatrix} (L_{aa} - L_{ac} - L_{ba} + L_{bc}) & (L_{ab} -- L_{ac} - L_{bb} + L_{bc}) & (L_{ar} - L_{br}) & 0 & 0 \\ (L_{ba} - L_{bc} - L_{ca} + L_{cc}) & (L_{bb} - L_{bc} - L_{cb} + L_{cc}) & (L_{br} - L_{cr}) & 0 & 0 \\ (L_{ra} - L_{rb}) & (L_{rb} - L_{rc}) & L_r & 0 & 0 \\ 0 & 0 & 0 & 1 & 0 \\ 0 & 0 & 0 & 0 & 1 \end{bmatrix} \tag{7.36}$$

$$A_2 = \begin{bmatrix} R_{as} & -R_{bs} & \omega\frac{\partial(L_{ar}-L_{br})}{\partial\theta} & 0 & 0 \\ R_{cs} & (R_{bs}+R_{cs}) & \omega\frac{\partial(L_{br}-L_{cr})}{\partial\theta} & 0 & 0 \\ \omega\frac{\partial(L_{ra}-L_{rc})}{\partial\theta} & \omega\frac{\partial(L_{rb}-L_{rc})}{\partial\theta} & R_r & 0 & 0 \\ 0 & 0 & 0 & 0 & 0 \\ 0 & 0 & 0 & -1 & 0 \end{bmatrix} \tag{7.37}$$

Additionally, the following assumptions simplify the solution of Equation (7.31).

1. Application of reciprocity theorem and symmetry considerations reduces the number of mutual inductance calculations, since all the magnetic and electric circuits are considered linear. For a healthy machine the stator phase to rotor loop inductances are also similar except for the phase shift. Also for multipolar machines all the inductances repeat after every 360° electrical.
2. The resistances and leakage inductances of the stator phases are considered identical. The same is true for rotor bars and end-ring segments. This is not true, however, in any practical motor. For example, the rotor bars, due to blow holes caused during the manufacturing process, will show variation in resistance.
3. Depending upon the solvers used, A_1 and parts of A_2 can be precomputed and stored in memory for different rotor positions. This way computation time can be saved, especially when each of the inductance computations involve several terms.
4. Since the rotor loop inductances and resistances are small, computational errors can be minimized by scaling them.

The mutual and magnetizing inductances used in the solution of Equation (7.31) can be computed by using either the winding function approach (WFA) or FE method. However, the FE method of computing the inductance is only good if stored inductance data is used, since it is very time consuming. The WFA can be used to compute inductances at every iteration if the computations are fast enough.

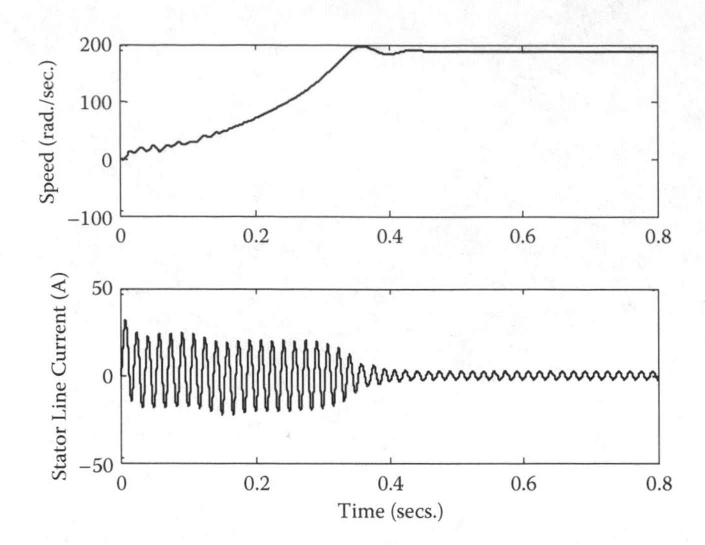

FIGURE 7.4
Simulated starting transients of an unloaded 4 pole, 3 hp induction motor. Speed (top), stator line current (bottom).

A few simulation results of an unskewed 3 hp, 4 pole, 28 rotor bar machine are presented by solving Equation (7.31) using MATLAB. The inductances have been recomputed at every iteration. Figure 7.4 shows the unloaded starting transient of the motor. The steady state current and speed under full load condition is shown in Figure 7.5. One cycle of rotor current is also shown in the

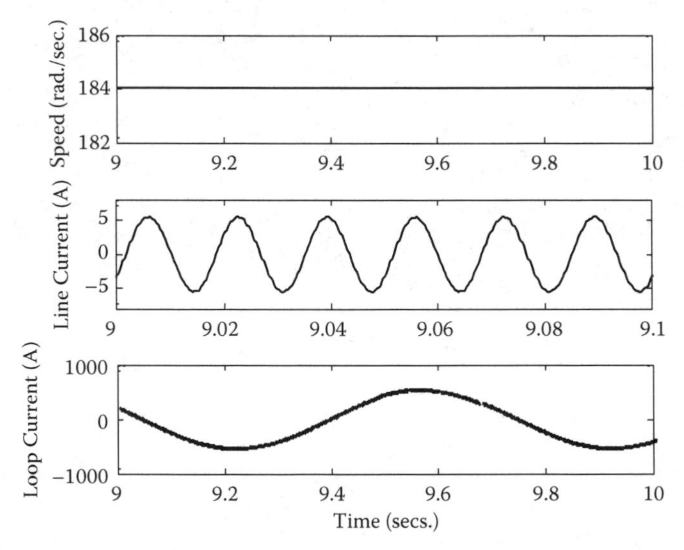

FIGURE 7.5
Simulated steady state performance of a loaded 4 pole, 3 hp induction motor. Speed (top), stator line current (middle), and rotor loop 1 current (bottom).

same figure. The computed current for the end-ring is zero under steady-state. The computation time for the unloaded motor was around 32 seconds (for 0.8 seconds simulation time), and 432 seconds (for 10.5 seconds of simulation time) on a 3.4 GHz, 1 GB, Pentium 4 machine running on Windows XP.

7.3 Model of Three-Phase Squirrel-Cage Induction Motor with Stator Inter-Turn Faults

7.3.1 Model without Saturation

The stator inter-turn fault can be modeled by considering another additional stator circuit [5,6]. This additional circuit f can be represented by the following equation

$$0 = R_{fs} i_{fs} + \frac{d\Lambda_{fs}}{dt} \quad (7.38)$$

with

$$\Lambda_f = L_{fs}\, \bar{i}_{fs} + L_{fr}\, \bar{i}_r \quad (7.39)$$

where

$$L_{fs} = [L_{ff} L_{fa}\; L_{fb}\; L_{fc}] \quad (7.40)$$

$$\bar{i}_s = [i_{fs} i_{as}\; i_{bs}\; i_{cs}]' \quad (7.41)$$

$$L_{fr} = [L_{fr_1}\; L_{fr_2} \cdot \cdots L_{fr_{n+l}}] \quad (7.42)$$

All the stator–stator and stator–rotor mutual inductances will have an extra term due to this shorted loop. Thus

$$L_{as} = [L_{af} L_{aa}\; L_{ab}\; L_{ac}] \quad (7.43)$$

$$L_{bs} = [L_{bf} L_{ba}\; L_{bb}\; L_{bc}] \quad (7.44)$$

$$L_{cs} = [L_{cf} L_{ca}\; L_{cb}\; L_{cc}] \quad (7.45)$$

Equations (7.16), (7.17), and (7.26) are now rewritten as

$$v_{ab} = v_a - v_b = R_{as}i_{as} - R_{bs}i_{bs} + (L_{af} - L_{bf})\frac{di_{fs}}{dt} + (L_{aa} - L_{ac} - L_{ba} + L_{bc})\frac{di_{as}}{dt} +$$

$$(L_{ab} - L_{ac} - L_{bb} + L_{bc})\frac{di_{bs}}{dt} + \omega\frac{\partial(L_{ar} - L_{br})}{\partial\theta}\bar{i}_r + (L_{ar} - L_{br})\frac{d\bar{i}_r}{dt} \tag{7.46}$$

$$v_{bc} = v_b - v_c = R_{cs}i_{as} + (R_{cs} + R_{bs})i_{bs} + (L_{bf} - L_{cf})\frac{di_{fs}}{dt} + (L_{ba} - L_{bc} - L_{ca} + L_{cc})\frac{di_{as}}{dt} +$$

$$(L_{bb} - L_{bc} - L_{cb} + L_{cc})\frac{di_{bs}}{dt} + \omega\frac{\partial(L_{br} - L_{cr})}{\partial\theta}\bar{i}_r + (L_{br} - L_{cr})\frac{d\bar{i}_r}{dt} \tag{7.47}$$

$$\bar{v}_r = R_r\bar{i}_r + L_{rf}\frac{di_{fs}}{dt} + (L_{rb} - L_{rc})\frac{di_{as}}{dt} + (L_{rb} - L_{rc})\frac{di_{bs}}{dt} + \omega\frac{\partial L_{rf}}{\partial\theta}i_{fs} +$$

$$\omega\frac{\partial(L_{ra} - L_{rc})}{\partial\theta}i_{as} + \omega\frac{\partial(L_{rb} - L_{rc})}{\partial\theta}i_{bs} + L_r\frac{di_r}{dt} \tag{7.48}$$

Also, here

$$L_{sr} = \begin{bmatrix} L_{fr} \\ L_{ar} \\ L_{br} \\ L_{cr} \end{bmatrix} \tag{7.49}$$

$$x = \left[i_{fs}\, i_{as}\, i_{bs}\, \bar{i}_r\, \omega\, \theta\right]^t \tag{7.50}$$

$$u = \left[0\; v_{ab}\; v_{bc}\; 0\cdots 0\; \frac{(T_m - T_l)}{J}\; 0\right]^t \tag{7.51}$$

$$A_1 = \begin{bmatrix} L_{ff} & (L_{fa} - L_{fc}) & (L_{fb} - L_{fc}) & L_{fr} & 0 & 0 \\ (L_{af} - L_{bf}) & (L_{aa} - L_{ac} - L_{ba} + L_{bc}) & (L_{ab} - L_{ac} - L_{bb} + L_{bc}) & (L_{ar} - L_{br}) & 0 & 0 \\ (L_{bf} - L_{cf}) & (L_{ba} - L_{bc} - L_{ca} + L_{cc}) & (L_{bb} - L_{bc} - L_{cb} + L_{cc}) & (L_{br} - L_{cr}) & 0 & 0 \\ L_{rf} & (L_{ra} - L_{rc}) & (L_{rb} - L_{rc}) & L_r & 0 & 0 \\ 0 & 0 & 0 & 0 & 1 & 0 \\ 0 & 0 & 0 & 0 & 0 & 1 \end{bmatrix} \tag{7.52}$$

$$A_2 = \begin{bmatrix} R_{fs} & 0 & 0 & \omega\frac{\partial L_{fr}}{\partial\theta} & 0 & 0 \\ 0 & R_{as} & -R_{bs} & \omega\frac{\partial(L_{ar}-L_{br})}{\partial\theta} & 0 & 0 \\ 0 & R_{cs} & (R_{bs}+R_{cs}) & \omega\frac{\partial(L_{br}-L_{cr})}{\partial\theta} & 0 & 0 \\ \omega\frac{\partial L_{rf}}{\partial\theta} & \omega\frac{\partial(L_{ra}-L_{rc})}{\partial\theta} & \omega\frac{\partial(L_{rb}-L_{rc})}{\partial\theta} & R_r & 0 & 0 \\ 0 & 0 & 0 & 0 & 0 & 0 \\ 0 & 0 & 0 & 0 & -1 & 0 \end{bmatrix} \tag{7.53}$$

It is assumed here that the faulty part of the winding belongs to the *a* phase of the motor. Hence R_{as} will not be equal to R_{bs} or R_{cs}. Also, the inductances involving the *a* phase or the faulty winding are computed differently. Using the subscript *h* for the original healthy winding

$$L_{aa} = L_{hh} - 2L_{hf} + L_{ff} \tag{7.54}$$

$$L_{af} = L_{hf} - L_{ff} \tag{7.55}$$

$$L_{ab} = L_{hb} - L_{fb} \tag{7.56}$$

$$L_{ac} = L_{hc} - L_{fc} \tag{7.57}$$

$$L_{ar} = L_{hr} - L_{fr} \tag{7.58}$$

Figure 7.6 and Figure 7.7 show the effect of shorting 5 turns out of 252 turns of the *a* phase of the 28 bar motor described in the previous section. The machine has been fully loaded. The top part of the plot shows the current in the faulty section of the winding. The bottom part of the plot also shows the effect of stator fault on the phase currents, which clearly become unbalanced. In this case the faulty phase has a maximum current followed by the *b* and the *c* phase. Figure 7.7 shows the rotor loop 1 current and the speed for the same faulty machine. The spikes in the loop current are the result of the loop picking up the faulty phase current as it passes by the faulty section of the winding. Because of the reverse rotating field (due to the unbalance produced by the fault) the speed also shows the presence of second harmonics. The unbalance in phase currents and speed can occur due to voltage or inherent constructional unbalance of the machine. Since there is no way to measure the current in the faulty section of the winding or the rotor loops,

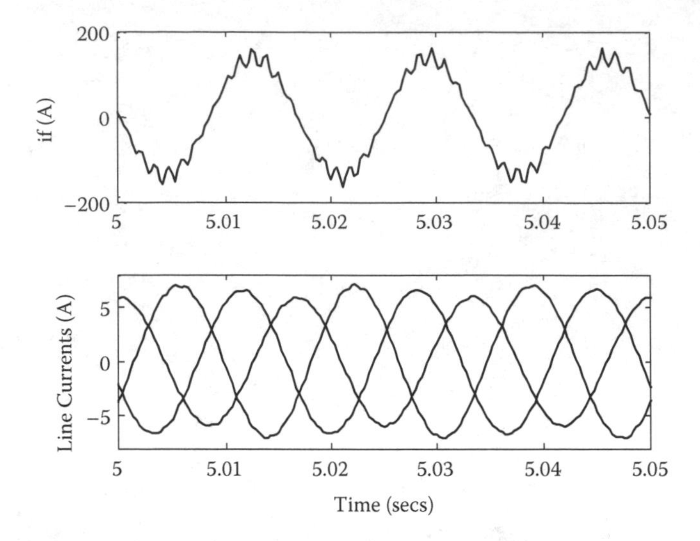

FIGURE 7.6
Simulated fault and line currents of a loaded 4 pole, 3 hp, 28 bar induction motor with stator fault. Current in the faulty part (top), the three phase currents (bottom).

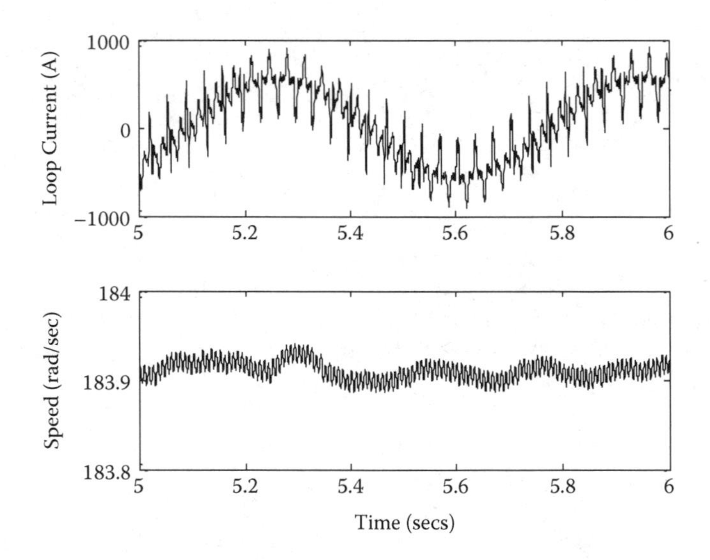

FIGURE 7.7
Simulated rotor loop current and speed of a loaded 4 pole, 3 hp, 28 bar induction motor with stator fault. Current in rotor loop 1 (top) and speed (bottom).

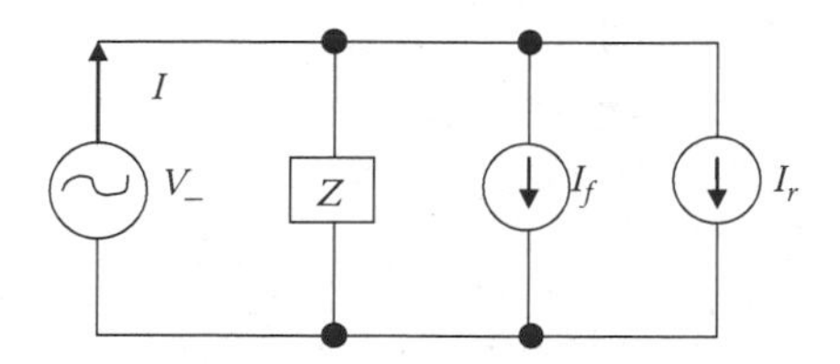

FIGURE 7.8
A simple model to diagnose stator faults.

stator faults are not easy to pick up. However due to heavy localized heating, the ground wall insulation may break down very soon (in a matter of a few seconds) leading to fire, explosion, and injury hazards. Hence stator faults have to be sensed very quickly.

Oftentimes, simpler models than the one described before have been used to detect stator faults on-line. One of these models used a simple negative sequence current-based approach [7]. It was shown that, provided the inter-turn short is confined to a small number of turns, the negative sequence impedance of the induction motor did not vary much. Since negative sequence current, I_- of a healthy motor is primarily dictated by its negative sequence impedance, Z_-, in conjunction with the negative sequence voltage, V_-, and the inherent machine unbalance and measurement inaccuracies, the negative sequence current due to these effects (V_-/Z_- and I_r) can be computed separately and subtracted from the total negative sequence current, I_- to compute I_f, the negative sequence current arising only out of the fault. Figure 7.8 illustrates the basic idea. Even single-turn dead bolt faults out of a total 648 turns/phases could be clearly identified using this method. However, the residual factor arising out of inherent machine unbalance and measurement inaccuracies is load dependent. Hence a look-up table has to be made for different load points.

It was later shown that the negative sequence current is dependent even on the positive sequence voltage [8]. This gives rise to nonzero off-diagonal coupling impedance terms between the sequence voltages and currents. Apart from measuring changes in the negative sequence impedance, changes in an off-diagonal term of the sequence component impedance matrix have also been utilized in order to detect stator inter-turn faults in the presence of inherent structural asymmetry and voltage unbalance [9]. The scheme also used a simple model for stator fault detection. A single-turn dead bolt fault out of a total 216 turns/phases can be detected. This scheme also requires storing data prior to motor operation or additional hardware.

7.3.2 Model with Saturation

The model derived in the earlier section [4] for induction motor did not include saturation. When saturation was included it opened up a new line of thought for fast and single-turn sensitive detection of stator fault in induction motors [10]. It is based on the 3rd harmonic components generated by the interaction of saturation-related permeance and magnetomotive force (MMF) harmonics. A line current 3rd harmonic (+3f) detecting scheme has been attempted for a direct torque control (DTC) induction machine drive to detect stator inter-turn faults [11]. It has been reported that a strong 3rd harmonic in the motor supply current has been introduced by the action of the torque and flux controllers. Unfortunately, the inherent asymmetry of the motor would also lead to the appearance of 3rd harmonic components in the supply current. Hence, this method is only recommended when a high degree of intrinsic asymmetry of the motor is not expected. Wu et al. validate a considerable increase of line current +3f components under fault conditions both numerically and experimentally [12].

Saturation-related specific air-gap permeance harmonics can be given as

$$P = P_m \cos(mp\theta - m\omega t) \tag{7.59}$$

Interaction of Equation (7.59) with reverse rotating MMF harmonics such as those given in Table 7.1 will result in a new series of flux density harmonics. Some specific flux density harmonics together with correlated principle time and permeance harmonics are listed in Table 7.2. It can be observed that the resulting pole pair numbers of $\pm 3\omega$ -related flux density harmonics have matching pole pair numbers with a standard three-phase integral slot winding, indicating that time harmonics similar to the ones shown in Table 7.2 are able to introduce both +3f and −3f line current harmonics. However, the introduction of these two components is following a different combination between MMF and permeance harmonics. Hence fault signatures will vary considerably between +3f and −3f. One example, according to Table 7.2, is that the fundamental reverse rotating field (given as $F_{1\text{-}}\cos(p\theta + \omega t)$) results in a +3f component only. It can be easily inferred that inherent structural asymmetry, magnetic anisotropy, and so on and supply unbalance will introduce both ±3f components although +3f, −3f arise from the interaction of different MMF

TABLE 7.1

*n*th MMF Space Harmonic Associated with *h*th Current Time Harmonic

	Order of Time Harmonic h		
Order of Space Harmonic N	**1**	**$6k-1$**	**$6k+1$**
1	+/−	−/+	+/−
$6s-1$	−/+	+/−	−/+
$6s+1$	+/−	−/+	+/−

Note: The signs before the slash (/) indicate those occurring under ideal condition; the signs after the slash (/) represent those occurring due to nonideal condition ($s, k = 1, 2, 3, \ldots$).

TABLE 7.2

MMF Harmonics, Air-Gap Permeance Harmonics, ±3f-Related Flux Densities, and ±3f Line Current Induced

MMF Harmonics	Permeance Harmonics	±3ω-Related Flux Density Harmonics	Line Current Harmonics
$F_{1-}\cos(p\theta + \omega t)$	$P_2\cos(2p\theta - 2\omega t)$	$B_{3+}\cos(p\theta - 3\omega t)$	+3f
$F_{1+}\cos(5p\theta - \omega t)$	$P_2\cos(2p\theta - 2\omega t)$	$B_{3+}\cos(7p\theta - 3\omega t)$	+3f
$F_{5-}\cos(5p\theta + 5\omega t)$	$P_8\cos(8p\theta - 8\omega t)$	$B_{3+}\cos(13p\theta - 3\omega t)$	+3f
$F_{5+}\cos(7p\theta - 5\omega t)$	$P_2\cos(2p\theta - 2\omega t)$	$B_{3+}\cos(5p\theta - 3\omega t)$	+3f
$F_{1+}\cos(5p\theta - \omega t)$	$P_4\cos(4p\theta - 4\omega t)$	$B_{3-}\cos(p\theta + 3\omega t)$	−3f
$F_{1-}\cos(7p\theta + \omega t)$	$P_2\cos(2p\theta - 2\omega t)$	$B_{3-}\cos(5p\theta + 3\omega t)$	−3f
$F_{5+}\cos(p\theta - 5\omega t)$	$P_2\cos(2p\theta - 2\omega t)$	$B_{3-}\cos(p\theta + 3\omega t)$	−3f
$F_{5-}\cos(5p\theta + 5\omega t)$	$P_2\cos(2p\theta - 2\omega t)$	$B_{3-}\cos(7p\theta + 3\omega t)$	−3f

and permeance harmonics. Some examples are given in Table 7.2. Also from the current spectrum shown later it can be observed that higher-order time harmonics (like 7th, 11th, etc.) are also capable of inducing ±3f components.

The preceding facts are clearly evident from the plots of the simulated motor as shown in Figure 7.9. The following cases have been simulated: (a) healthy saturated machine with balanced three-phase supply, (b) healthy saturated machine with unbalanced three-phase supply, and (c) faulty saturated machine with single-turn fault but with balanced three-phase supply. In all these cases, the supply voltage was harmonics-free. A substantial increase of +3f component can be seen both in the presence of voltage unbalance as well as inter-turn fault. The −3f component, though absent in case of both (a) and (b), clearly shows up in (c) due to the presence of a fault. This implies that different metrics have to be used for the two components.

Following this logic it can be inferred that current harmonics instead of voltage can be directly used to detect the faults, avoiding expensive voltage sensors. Although Table 7.2 lists many MMF harmonics, only the ones with significant amplitude will influence detection. For example, the magnitude spectra in Figure 7.10 reveal that f, −f, −5f, +5f, +7f, and −11f have magnitudes of −50 dB or above for an experimental machine fed from utility supply. Hence they can be selected to determine least-square-based estimates of the coefficients associated with each harmonic component. The number of coefficients should be optimized such that the best detection is possible without increased computational penalty. Hence six complex coefficients ($\bar{k}_{I0+}$-$\bar{k}_{I5-}$) or ($\bar{k}_{I0-}$-$\bar{k}_{I5-}$) have been used. The expressions for computing estimates (denoted by the subscript *e*) are given as

$$\bar{I}_{I3+(e)} = \bar{k}_{I0+}\bar{I}_{1+} + \bar{k}_{I1+}\bar{I}_{1-} + \bar{k}_{I2+}\bar{I}_{5+} + \bar{k}_{I3+}\bar{I}_{5-} + \bar{k}_{I4+}\bar{I}_{7+} + \bar{k}_{I5+}\bar{I}_{11-} \quad (7.60)$$

$$\bar{I}_{I3-(e)} = \bar{k}_{I0-}\bar{I}_{1+} + \bar{k}_{I1-}\bar{I}_{1-} + \bar{k}_{I2-}\bar{I}_{5+} + \bar{k}_{I3-}\bar{I}_{5-} + \bar{k}_{I4-}\bar{I}_{7+} + \bar{k}_{I5-}\bar{I}_{11-} \quad (7.61)$$

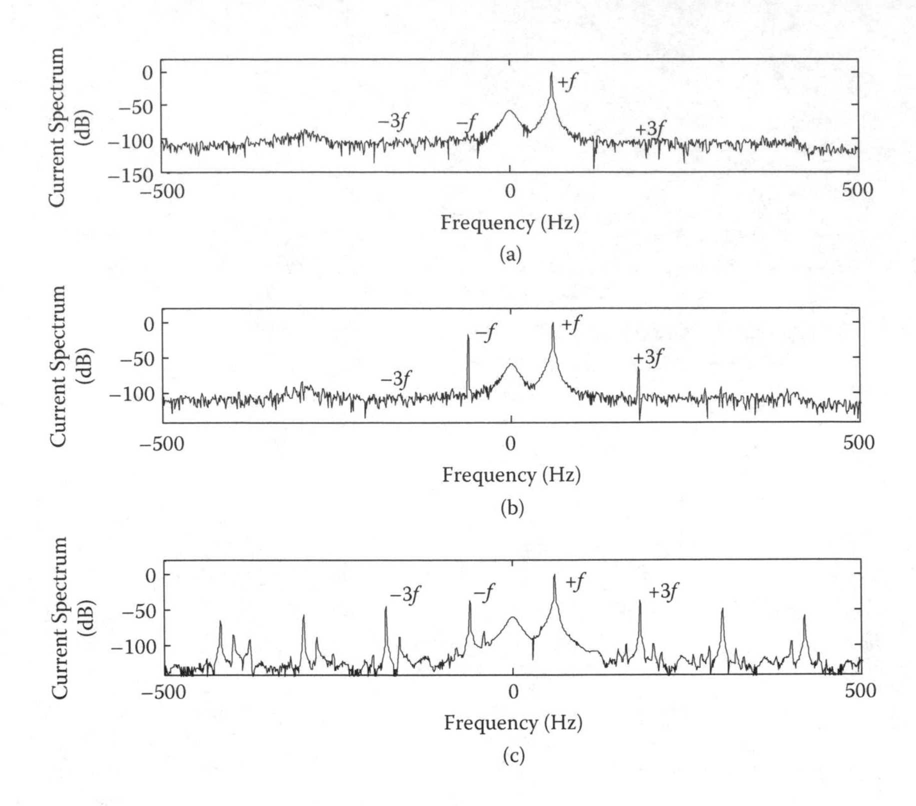

FIGURE 7.9
Normalized, simulated, magnitude spectrum of line current space vector with full load under (a) balanced, (b) unbalanced, and (c) single-turn fault with balanced voltage.

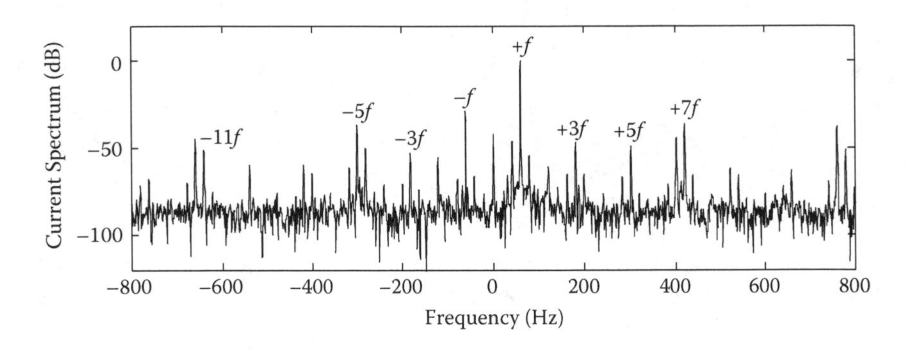

FIGURE 7.10
Experimentally computed normalized, magnitude spectrum of line current space vector under full load condition.

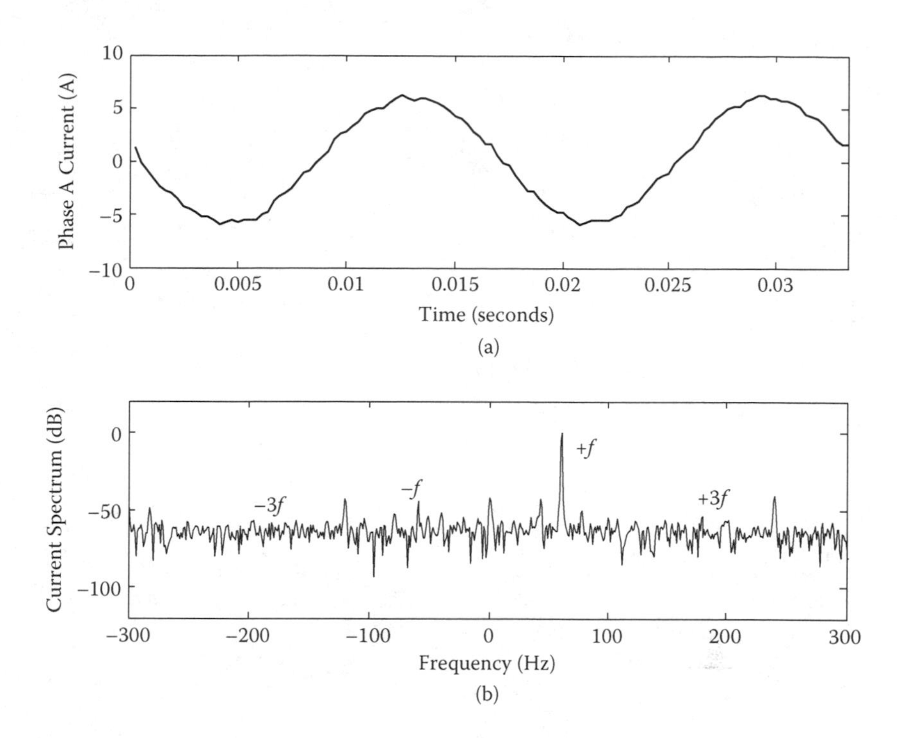

FIGURE 7.11
Full load (a) two cycles of healthy phase *a* current; (b) normalized, magnitude spectrum of line current space vector of healthy machine at 60 Hz under.

Nine data sets were used to obtain the six coefficients. Once the coefficients have been computed, the final fault signatures (residues) are obtained by subtracting estimates from measured quantities (denoted by the subscript *m*) given as

$$\mathrm{FSI}_{3+} = \left|\overline{I}_{3+(m)} - \overline{I}_{3+(e)}\right| \text{ or } \mathrm{FSI}_{3-} = \left|\overline{I}_{3-(m)} - \overline{I}_{3-(e)}\right| \tag{7.62}$$

Surprisingly, when the same machine was fed by an inverter, the spectra looked a lot cleaner (Figure 7.11) and only four coefficients were required. Also, the level of voltage or structural unbalance was very low as can be seen from the low level of –f component in the line current.

Tables 7.3 to 7.6 list some of the residues as computed by Equation (7.61). The sensitivity is very good. The residue under faulty cases is always more than 10 dB below the corresponding healthy balanced and healthy unbalanced residues.

TABLE 7.3

fsi3+ of Line-Fed, Star-Connected Machine

	Normalized +3f Signature (dB) under Different Operating Conditions						
Rotating Speed (rpm)	**HB**	**HU**	**T1**	**T2**	**T3**	**T4**	**T5**
1799	–61.6	–44.3	–28.1	–28.1	–25.6	–25.2	–23.2
1790	–52.0	–40.5	–15.1	–15.3	–14.9	–13.0	–12.0
1780	–48.9	–35.3	–16.5	–16.5	–18.3	–16.4	–18.4
1770	–51.7	–33.1	–20.0	–18.4	–17.1	–17.3	–16.2
1760	–54.7	–32.1	–17.3	–19.1	–17.5	–17.6	–19.2

TABLE 7.4

fsi3– of Line-Fed, Star-Connected Machine

	Normalized –3f Signature (dB) under Different Operating Conditions						
Rotating Speed (rpm)	**HB**	**HU**	**T1**	**T2**	**T3**	**T4**	**T5**
1799	–64.4	–47.1	–33.9	–33.5	–32.4	–32.3	–30.1
1790	–59.9	–48.9	–33.5	–31.8	–29.6	–26.2	–25.4
1780	–62.0	–52.8	–36.6	–36.2	–40.2	–35.7	–40.9
1770	–64.2	–52.8	–35.5	–32.4	–30.5	–30.2	–29.4
1760	–68.1	–46.0	–35.7	–36.5	–35.1	–35.0	–34.4

TABLE 7.5

fsi3+ of Inverter-Fed Machine (60 Hz)

	Normalized +3f Signature (dB) under Different Operating Conditions					
Rotating Speed (rpm)	**HB**	**T1**	**T2**	**T3**	**T4**	**T5**
1799	–58.7	–35.6	–38.0	–30.1	–30.0	–31.2
1790	–53.7	–38.4	–33.0	–29.2	–25.9	–26.4
1780	–61.5	–38.7	–36.0	–34.8	–30.9	–29.7
1770	–63.7	–46.7	–41.1	–45.9	–35.4	–32.7
1760	–59.9	–37.7	–37.8	–34.7	–31.3	–32.8

TABLE 7.6

fsi3– of Inverter-Fed Machine (60 Hz)

	Normalized –3f Signature (dB) under Different Operating Conditions					
Rotating Speed (rpm)	**HB**	**T1**	**T2**	**T3**	**T4**	**T5**
1799	–65.9	–43.9	–44.5	–41.3	–37.9	–39.7
1790	–65.9	–49.4	–43.8	–36.9	–35.6	–32.6
1780	–73.7	–54.0	–47.5	–39.1	–37.1	–35.7
1770	–76.0	–50.5	–48.1	–47.7	–41.5	–37.6
1760	–74.5	–50.7	–51.3	–44.4	–42.7	–38.7

7.4 Model of Squirrel-Cage Induction Motor with Incipient Broken Rotor Bar and End-Ring Faults

One way to simulate a broken rotor bar or end-ring fault is by simply removing that element and rewriting the circuit equation by eliminating and redefining loops (see Equation 7.26) [13]. This is, however, not exactly how a standard squirrel-cage bar breaks or a broken rotor bar effect is felt in a real machine. In most of the small and medium-sized squirrel-cage motors, the rotors are fabricated by using die-cast aluminum, molded to form an integral block with the rotor laminations. Thus there is no insulation between the bar/ end-ring and the core. As a result, considerable inter-bar current exists even though a rotor bar may be completely broken. Also the bar breakage starts as a crack rather than a complete disconnection between the two section of a bar. Even for fabricated rotors, where aluminum or copper bars are inserted into the rotor laminations with end-rings brazed, welded or molded onto the bars, complete bar, or end-ring removal in the simulation may not correctly describe the actual scenario. This is because of the fact that even the bars of a fabricated rotor are not always insulated from the core.

The arguments presented in the previous paragraph show that it is best to simulate an incipient broken rotor bar fault as an increase in resistance of the bar or, to be more precise, the loop resistances used in Equation (7.26). Thus to simulate a partially broken rotor bar, the matrix R_r as given in Equation (7.20) is changed in such a way that only two of the loop equations are affected. This is because a bar is included in two loops, whereas a broken end-ring segment is included in one loop only. In general, for m number of partially broken rotor bars or end-ring segments $m+1$ or m loop equations are affected, respectively. Rotor-stator or rotor-rotor mutual or self-inductances remain unchanged in either case.

For m fully broken rotor bars or end-ring segments, m loop equations are removed. Rotor–stator or rotor–rotor mutual or self-inductances are however changed only when the rotor bars are fully broken.

Figure 7.12 shows the effect of broken rotor bars with resistances of four consecutive bars increased from 50 μΩ to 1 mΩ in a 44 bar, 4 pole, 3 hp

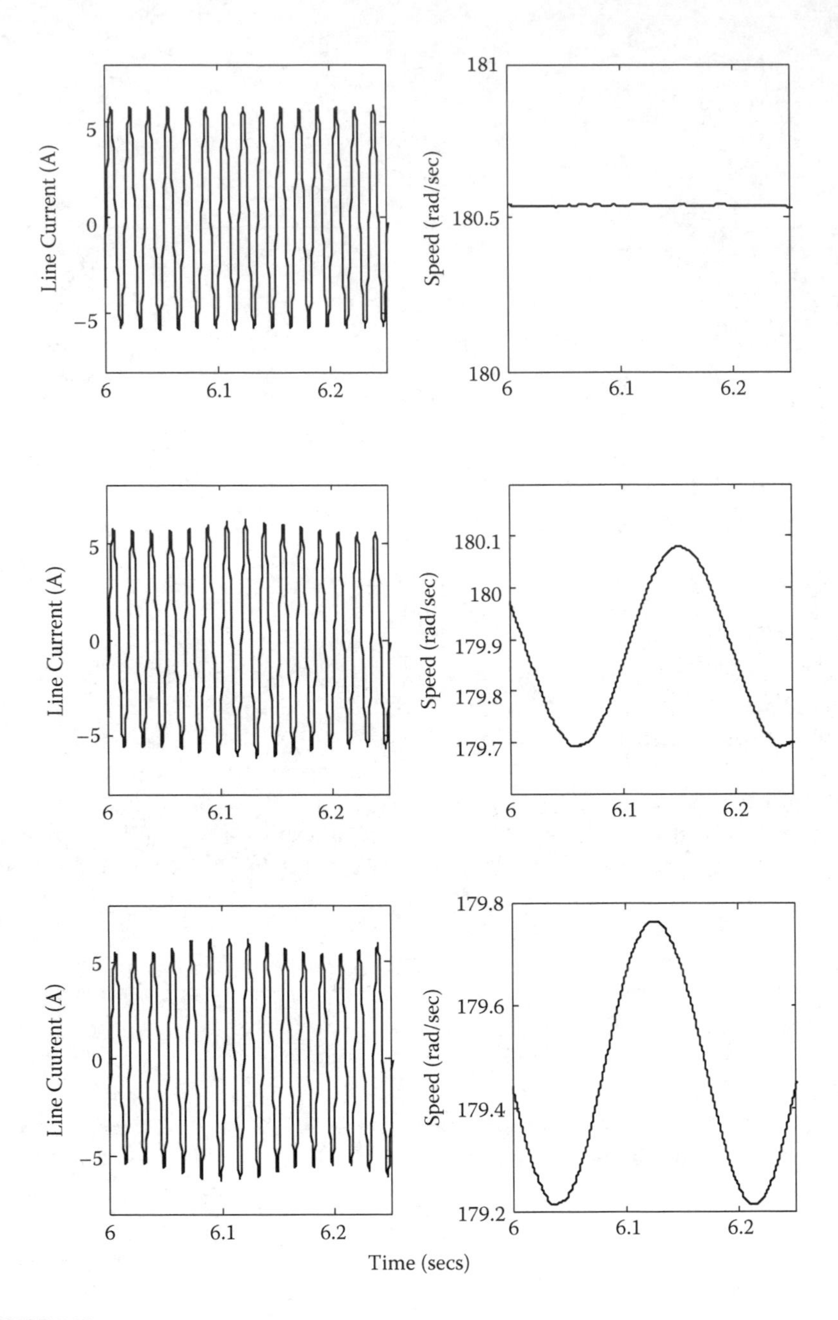

FIGURE 7.12
Simulated line currents and speed of a loaded 4 pole, 3 hp, 44 bar induction motor with and without broken rotor bar fault. Healthy (top), with four partially broken rotor bars (middle), and two partially broken end-rings (bottom).

induction motor. Compared to the healthy condition the increased oscillations in current and speed are quite evident. Similar effects were observed when resistances of two end-rings were increased from 2.7 μΩ to 25 mΩ.

7.5 Model of Squirrel-Cage Induction Motor with Eccentricity Faults

Eccentricity faults can be modeled by modifying the specific permeance function of a machine [14]. This would essentially mean changing the air-gap function of a machine and then calculating the changes in inductances. The best way to do this would be to probably compute all the inductances using FE methods, storing them with sufficiently fine resolution, and then modeling the motor in a manner similar to the one described in Section 7.2. FE techniques though accurate are very time consuming.

As shown earlier in Chapter 3, the modified winding function approach can also be used to compute magnetizing and mutual inductances under eccentric conditions. This can be done by writing the air-gap as the following:

$$g_e(\phi,\theta_r) = g_0 - a_1\cos\phi - a_2\cos(\phi-\theta) \quad (7.63)$$

where a_1, a_2 are the amount of static and dynamic eccentricity respectively, g_0 the average air-gap, and ϕ a particular position along the stator inner surface. Then, the inverse air-gap function, $g_e^{-1}(\phi,\theta)$, can be written as

$$g_e^{-1}(\phi,\theta) = \frac{1}{g_0(1-a_3\cos(\phi-\theta_1))} \quad (7.64)$$

with

$$a_3 = \sqrt{a_1^2 + 2a_1a_2\cos\theta + a_2^2},\ \theta_1 = \arctan\left(\frac{a_2\sin\theta}{a_1 + a_2\cos\theta}\right) \quad (7.65)$$

The inverse air-gap function can be approximately expressed as

$$g_e^{-1}(\varphi,\theta_1) \approx A_1 + A_2\cos(\varphi,\theta_1) \quad (7.66)$$

where

$$A_1 = \frac{1}{g_0\sqrt{1-a_3^2}},\ A_2 = \frac{2}{g_0\sqrt{1-a_3^2}}\left(\frac{1-\sqrt{1-a_3^2}}{a_3}\right), \quad (7.67)$$

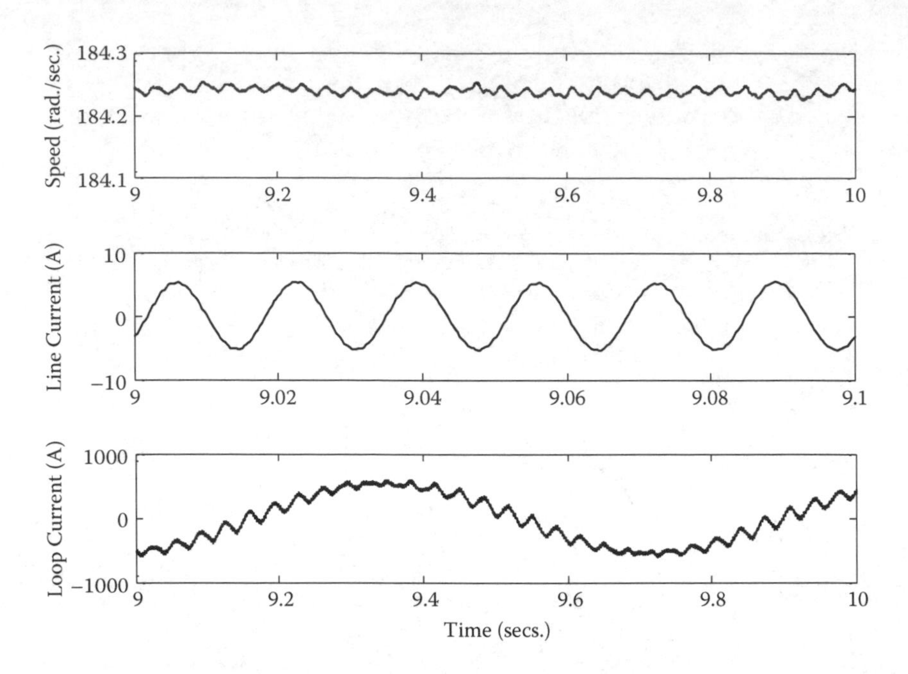

FIGURE 7.13
Simulated loaded 4 pole, 3 hp, induction motor with 41% static and 21% dynamic eccentricity. Speed (top), stator line current (middle), and rotor loop 1 current (bottom).

However, unlike the healthy machine described in Section 7.2, for a mixed eccentric machine, $\frac{\partial L_{ss}}{\partial \theta}, \frac{\partial L_{rr}}{\partial \theta}$ are nonzero quantities in the electromagnetic torque expression given by

$$T_m = 0.5\left[I_s^t \frac{\partial L_{ss}}{\partial \theta} I_s + I_s^t \frac{\partial L_{sr}}{\partial \theta} I_r + I_r^t \frac{\partial L_{rs}}{\partial \theta} I_s + I_r^t \frac{\partial L_{rr}}{\partial \theta} I_r \right] \tag{7.68}$$

Modeling of pure static eccentricity or pure dynamic eccentricity can be treated as special cases of mixed eccentricity. For pure static eccentricity a_2 equals zero and for pure dynamic eccentricity a_1 equals zero in Equation (7.63) and Equation (7.64). With static eccentricity, the torque expression (7.68) will not have $\frac{\partial L_{ss}}{\partial \theta}$ terms, whereas with dynamic eccentricity $\frac{\partial L_{rr}}{\partial \theta}$ terms will be absent in expression (7.68).

Figure 7.13 shows the plots of line speed, current, and rotor loop current of the 28 bar induction motor, whose simulation results under healthy conditions have been shown in Section 7.2, with 41% static eccentricity and 21% dynamic eccentricity. Comparison of Figure 7.13 with Figure 7.4 clearly shows that the mixed eccentric machine produces very distinct signatures. These signatures in frequency domain will be clearly able to diagnose eccentricity faults. Sometimes, however, depending on the motor pole pairs and number of rotor slots, these signatures may not be present in the time domain, making fault diagnosis very difficult. The frequency-domain-based signatures was discussed in Chapter 6.

Fault models with inclined static eccentricity have also been developed. The way the mutual and magnetizing inductances are computed is only different. The modeling equations are the same as those for the uniform static eccentric machines.

7.6 Model of a Synchronous Reluctance Motor with Stator Fault

A synchronous reluctance motor is essentially a synchronous motor that derives all of its power on rotor saliency since the field winding is absent. However, it has damper windings to provide stability of operation under various perturbations and transients. The motor is normally used for low power applications that require constant speed operation irrespective of the load torque. Additionally, the modeling of this machine can be treated as the first step for a synchronous machine. It also gives the option to isolate the effect of the damper winding from that of the field winding and can provide crucial insight in detecting various types of faults [15].

The modeling of this machine is similar to an induction machine, except for the fact that (1) the damper bars exist only on the pole faces of the machine and (2) the air-gap or inverse air-gap function has saliency. The former can be incorporated by removing the loops that exist in the inter-polar gap, in a way similar to that of an induction machine with broken rotor bars. The latter can be incorporated by modeling the machine like the saturated induction motor, with the inverse air-gap function.

$$g_s^{-1}(\phi,\theta) \approx \frac{1}{g_0}\left[1+\sum_{k=1,2,3..}^{\infty} a_{gk} Cos\{2kp(\phi-\theta)\}\right] \tag{7.69}$$

Figure 7.14 shows the simulated start-up and loading condition of the motor. The stator fault for this motor has been simulated in a similar way to the induction motor. There is a clear increase in the phase current of the motor with a four-turn fault as can be seen in Figure 7.15. In an actual machine a similar increase was seen (Figure 7.16), confirming the utility of the model developed for the motor.

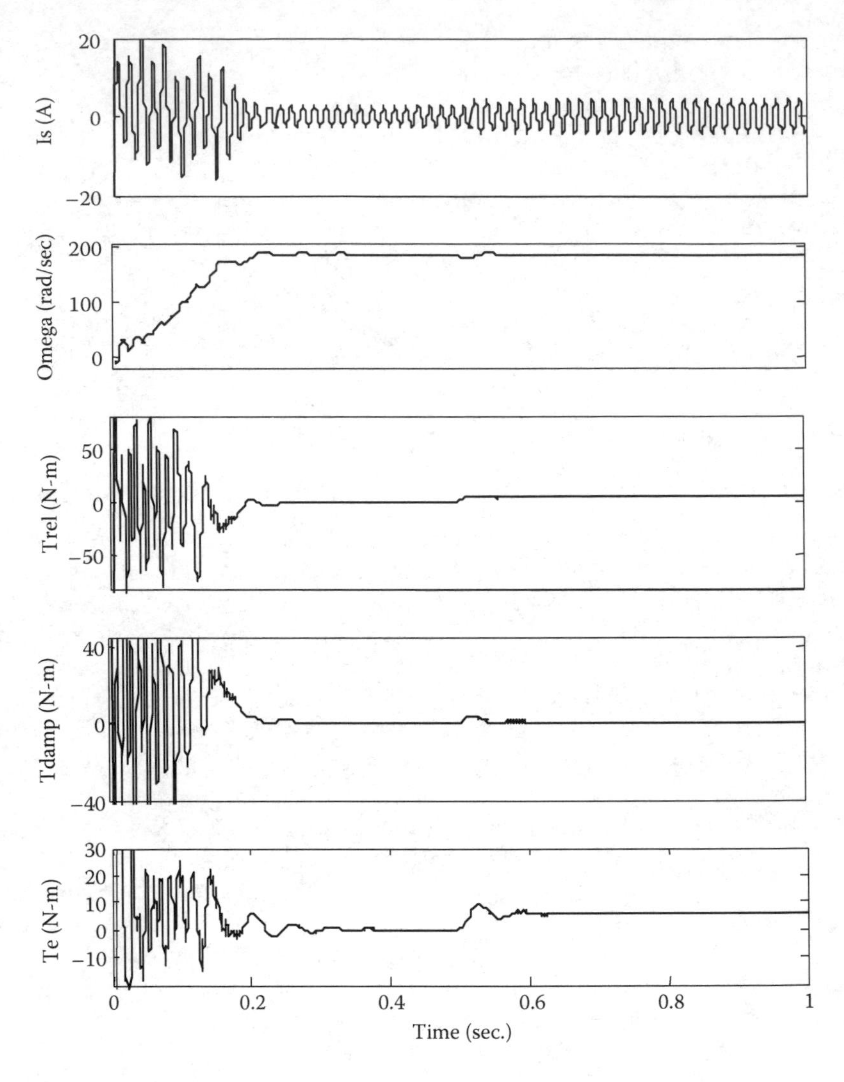

FIGURE 7.14
From top to bottom, simulated stator current (Is), speed (Omega), reluctance torque (T_{rel}), damping torque (T_{damp}), and total torque (T_e).

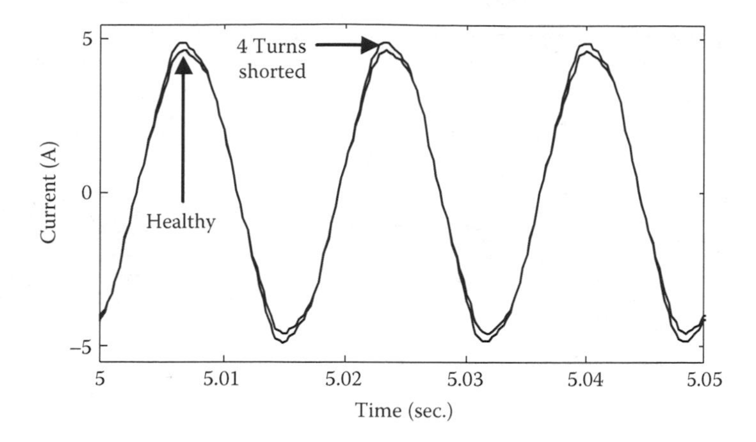

FIGURE 7.15
Simulated currents of stator phase *a*.

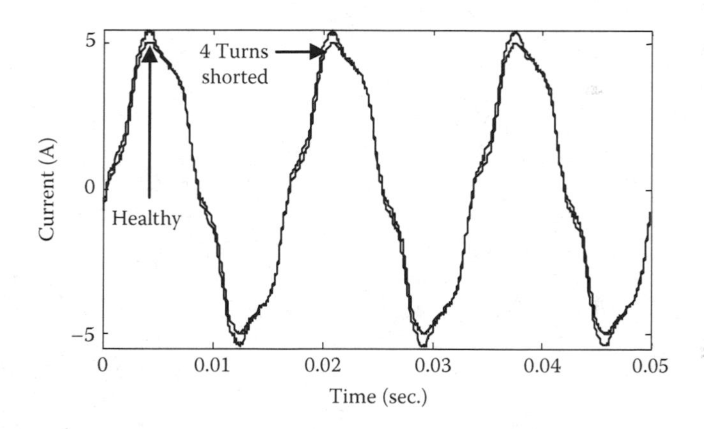

FIGURE 7.16
Experimental currents of stator phase *a*.

7.7 Model of a Salient Pole Synchronous Motor with Dynamic Eccentricity Faults

The dynamic eccentricity faults for a synchronous machine can be modeled in the same line as an induction motor [16,17]. However, unlike the induction machine, a salient pole synchronous motor does not have a smooth air-gap.

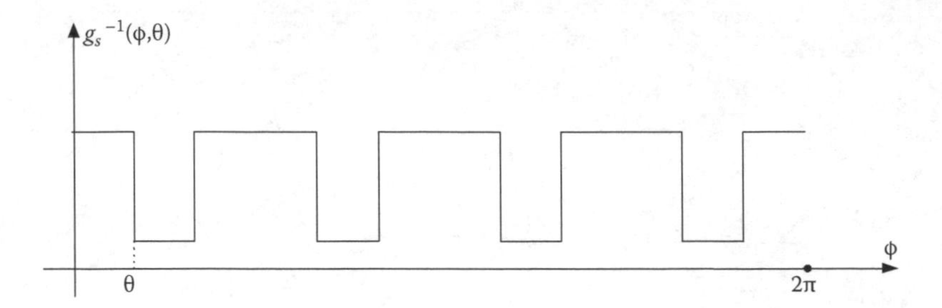

FIGURE 7.17
Inverse gap function in case of symmetric rotor.

The harmonic contents of the inverse air-gap shows only selective space harmonics so that it is symmetric. This is given by Equation (7.69). With dynamic eccentricity, all possible harmonics will be present in the inverse air-gap function. It can be therefore expressed as

$$g_{sd}^{-1}(\phi,\theta) \approx \frac{1}{g_o'}\left[1+\sum_{k=1,2,3..}^{\infty} a_{gkd} Cos\{k(\phi-\theta)\}\right] \tag{7.70}$$

The symmetric and asymmetric air-gaps are as shown in Figure 7.17 and Figure 7.18, respectively. Also in this case the damper bars are neglected and the field has got a single winding with a four-pole structure with a direct current voltage applied to it. The modeled machine showed increase in the 17th and 19th time harmonics in the presence of dynamic eccentricity. Experimental results with an unbalanced disc also showed a similar increase.

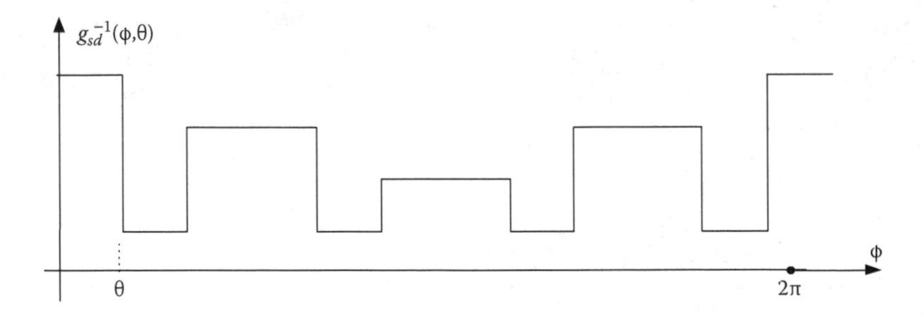

FIGURE 7.18
Inverse gap function in case of asymmetric rotor.

References

[1] K. Kim and A. Parlos, "Induction motor fault diagnosis based on neuropredictors and wavelet signal processing," *IEEE/ASME Transactions on Mechatronics*, vol. 7, no. 2, pp. 201–219, June 2002.

[2] F. Fischer, H.J. Nern, L. Lahtchev, and H.A. Nour Eldin, "Explicit modeling of the stator winding bar water cooling for model-based fault diagnosis of turbogenerators with experimental verification," *Proceedings of the 3rd IEEE Conference on Control Applications*, pp. 1403–1408, August 1994.

[3] A.L. Dexter, "Fuzzy model based fault diagnosis," *IEE Proceedings on Control Theory Application*, vol. 142, no. 6, pp. 545–550, November 1995.

[4] S. Nandi, "Condition Monitoring for Fault Diagnosis of Induction Motors," PhD dissertation, Texas A&M University, May 2000.

[5] S. Nandi and H.A. Toliyat, "A novel frequency domain-based technique to detect stator inter-turn faults in induction machines using stator induced voltages after switch-off," *IEEE Transactions on Industry Applications*, vol. 38, no. 1, pp. 101–109, January/February 2002.

[6] S. Nandi, "Detection of stator faults in induction machines using residual saturation harmonics," *IEEE IEMDC '05*, pp. 256–263, May 15–18, 2005.

[7] G.B. Kliman, W.J. Premerlani, R.A. Keogl, and D. Hoeweler, "A new approach to on-line turn fault detection in AC motors," *Proceedings of IEEE-IAS 1999 Annual Meeting Conference*, pp. 687–693, October 6–10, 1996.

[8] R.M. Tallam, T.G. Habetler, and R.G. Harley, "Transient model for induction machines with stator winding turn faults," *IEEE Transactions on Industry Applications*, vol. 38, no. 3, pp. 632–637, May/June 2002.

[9] S.B. Lee, R.M. Tallam, and T.G. Habetler, "A robust on-line turn-fault detection technique for induction machines based on monitoring the sequence component impedance matrix," *IEEE Transactions on Power Electronics*, vol. 18, no. 3, pp. 865–872, May 2003.

[10] S. Nandi, "A detailed model of induction machines with saturation extendable for fault analysis," *IEEE Transactions on Industry Applications*, vol. 40, no. 5, pp. 1302–1309, September/October 2004.

[11] S.M.A Cruz and A.J.M. Cardoso, "Diagnosis of stator inter-turn short circuits in DTC induction motor drives," *IEEE Transactions on Industry Applications*, vol. 40, no. 5, pp. 1349–1360, September/October 2004.

[12] S. Nandi and Q. Wu, "Fast single-turn sensitive stator inter-turn fault detection of induction machines based on positive and negative sequence third harmonic components of line currents," *IEEE Transactions on Industry Applications*, vol. 46, no. 3, pp. 974–983, May/June 2010.

[13] S. Nandi, R. Bharadwaj, H.A. Toliyat, and A.G. Parlos, "Study of three-phase induction motors with incipient cage faults under different supply conditions," *Proceedings of IEEE-IAS 1999 Annual Meeting Conference*, pp. 1922–1928, October 1999.

[14] S. Nandi, R. Bharadwaj, and H.A. Toliyat, "Performance analysis of a three-phase induction motor under incipient mixed eccentricity condition," *IEEE Transactions on Energy Conversion*, vol. 17, no. 3, pp. 392–399, September 2002.

[15] P. Neti and S. Nandi, "Stator inter-turn fault analysis of reluctance synchronous motor," *Proceedings of IEEE CCECE Conference*, pp. 1283–1286, May 2005.

[16] N.A. Al-Nuaim and H.A. Toliyat, "A novel method for modeling dynamic air-gap eccentricity in synchronous machines based on modified winding function theory," *IEEE Transactions on Energy Conversion*, vol. 13, no. 2, pp. 156–162, June 1998.

[17] H.A. Toliyat and N.A. Al-Nuaim, "Simulation and detection of dynamic air-gap eccentricity in salient-pole synchronous machines," *IEEE Transactions on Industry Applications*, vol. 13, no. 4, pp. 86–93, January/February 1999.

8

Application of Pattern Recognition to Fault Diagnosis

Masoud Hajiaghajani, Ph.D

8.1 Introduction

The problem of fault diagnosis can be viewed as a classification problem if a proper selection of features and classifier is chosen. To begin with, we introduce the pattern recognition system structure and then the classifier design problem will be addressed for our fault diagnosis system. The presented technique will be used for both alternating current (AC) and direct current (DC) machines fault detection.

Many data-driven, knowledge-based, and analytical approaches incorporate pattern-based techniques to some extent. Pattern-based methods generally consist of templates or patterns distinguishing acceptable and unacceptable operations that are then compared to the system observations to determine whether a fault has occurred. These templates or patterns may be determined by performance specifications, by past observations of faulty operations, by expert knowledge, or even from analysis or simulation of a system model. Once trained, the system is able to rapidly recognize pattern similarities and classify new observations accordingly. The primary disadvantage, however, is that the success of the fault detection and diagnosis is strongly dependent upon the initial training data. The volume of training data required may be extensive, and only faults represented in the training data can be diagnosed. Observations that are significantly different from the training data can be incorrectly diagnosed.

A pattern recognition system contains three parts: a transducer, a feature extractor, and a classifier. The transducer senses the input and converts it into a form suitable for machine processing. The feature extractor extracts presumably relevant information from the input data. The classifier uses information to assign the input data to one of a finite number of categories. A simple block diagram of a pattern recognition system for the problem of fault diagnosis is shown in Figure 8.1. In the next section, the fundamentals of Bayesian decision theory are introduced and shown how it can be

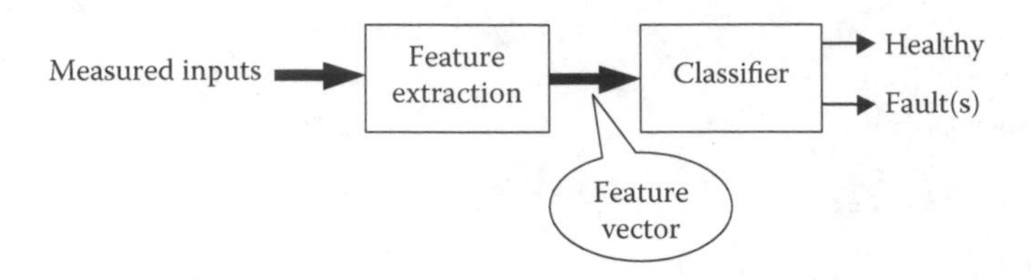

FIGURE 8.1
Pattern recognition approach for the fault diagnosis problem.

viewed as being simply a formalization of common-sense procedures. This theory is used for the design of our classifier. Feature extraction and feature assignment is covered afterward, which is based on the result of an analytical study presented in the previous chapters. Implementation on the experimental data is covered at the end of the chapter.

8.2 Bayesian Theory and Classifier Design

Bayesian decision theory is a fundamental statistical approach to the problem of pattern classification. This approach is based on quantifying the trade-offs between various classification decisions using probability and the costs that accompany such decisions. It makes the assumption that the decision problem is posed in probabilistic terms and that all of the relevant probability values are known.

Let's start with a simple example of designing a classifier to separate two kinds of car: compact and SUV. Suppose that an observer watching cars arrive along the road finds it hard to predict what type will come next and that the sequence of types of cars appears to be random. We let ω denote the *class* of cars, with $\omega = \omega_1$ for the class of compact cars and $\omega = \omega_2$ for SUV. Because the condition of the class is so unpredictable, we consider ω to be a variable that must be described. If the observation recorded as many compact cars as SUVs, we would say that the next car is equally likely to be a compact car or SUV. More generally, we assume that there is *a priori probability* $P(\omega_1)$ that the next car is a compact car, and some prior probability $P(\omega_2)$ that it is a SUV. If we assume there are no other types of cars relevant here, then $P(\omega_1)$ and $P(\omega_2)$ sum to one. These prior probabilities reflect our prior knowledge of how likely we are to observe a compact car or a SUV before the car actually appears. It might, for instance, depend upon the time of year or the neighborhood.

Suppose that we want to make a decision about the type of car that will appear next without being allowed to see it. Also assume that any incorrect classification entails the same cost or consequence, and that the only information we are allowed to use is the value of the prior probabilities. If a decision must be made with so little information, it seems logical to use the following *decision rule*:

$$\text{Decide } \omega_1 \text{ if } P(\omega_1) > P(\omega_2)\text{; otherwise decide } \omega_2 \tag{8.1}$$

How well it works depends upon the values of the prior probabilities. If $P(\omega_1)$ is much greater than $P(\omega_2)$, our decision in favor of ω_1 will be right most of the time. If $P(\omega_1) = P(\omega_2)$, we have only a fifty–fifty chance of being right. In general, the probability of error is the smaller of $P(\omega_1)$ and $P(\omega_2)$.

To present this theory in terms of mathematical language, let $\Omega = \{\omega_1, \omega_2, \ldots, \omega_s\}$ be the finite set of classes and $P(\omega_i)$ be the probability of each class. Having an observed vector x (or feature vector), the Bayesian theory says

$$P(\omega_i|\underline{x}) = \frac{p(\underline{x}|\omega_i)P(\omega_i)}{\sum_j p(\underline{x}|\omega_j)P(\omega_j)} \tag{8.2}$$

where $p(\underline{x}|\omega_i)$ is the class-conditional probability density function for $\underline{x}$; that is, the probability density function for $\underline{x}$ given that the class is ω_i. $P(\omega_i|\underline{x})$ is the probability of selected class (ω_i), given the feature vector $\underline{x}$. The strength of the preceding equation is that it relates our observation and priori probability, $p(\underline{x}|\omega_i)$, to a posteriori probability, $P(\omega_i|\underline{x})$ as is shown in Figure 8.2 and Figure 8.3.

If we have an observation x for which $P(\omega_1|\underline{x})$ is greater than $P(\omega_2|\underline{x})$, we would naturally be inclined to decide that the true class is ω_1. Similarly, if $P(\omega_2|\underline{x})$ is greater than $P(\omega_1|\underline{x})$, we would be inclined to choose ω_2. To justify

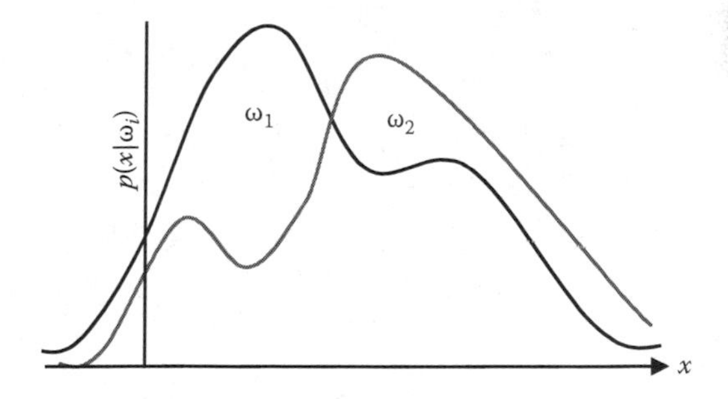

FIGURE 8.2
The probability density of measuring a particular feature value x given the pattern is in category ω. If x represents the length of a car, the two curves might describe the difference in length of populations of two types of cars. Density functions are normalized, and thus the area under each curve is 1.0.

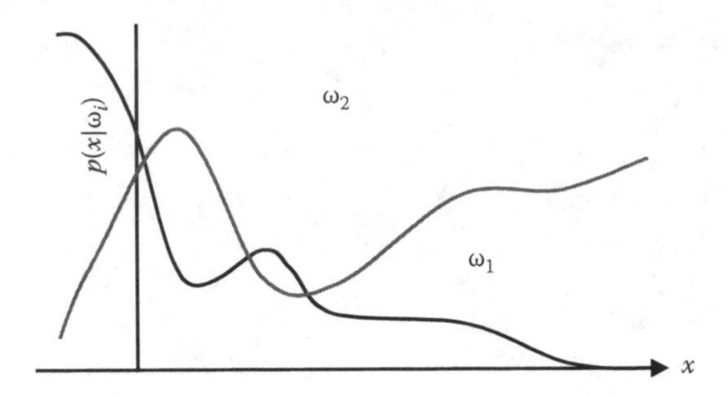

FIGURE 8.3
Posterior probabilities for the particular priors $P(\omega_1) = 2/3$ and $P(\omega_2) = 1/3$ for the class-conditional probability densities shown in Figure 8.2. Thus in this case, given that a pattern is measured to have feature value $x = 14$, the probability it is in category ω_2 is roughly 0.08, and that it is in ω_1 is 0.92. At every x, the posteriors sum to 1.0.

this decision procedure, let us calculate the probability of error whenever we make a decision. Whenever we observe a particular x,

$$P(error|x) = \begin{cases} P(\omega_1|x) & if \quad we \quad decide \quad \omega_2 \\ P(\omega_2|x) & if \quad we \quad decide \quad \omega_1 \end{cases} \tag{8.3}$$

Clearly, for a given x we can minimize the probability of error by deciding ω_1 if $P(\omega_1|\underline{x}) > P(\omega_2|\underline{x})$ and ω_2 otherwise. Of course, we may never observe exactly the same value of x twice. Will this rule minimize the average probability of error? Yes, because the average probability of error is given by

$$P(error) = \int_{-\infty}^{\infty} P(error, x)dx = \int_{-\infty}^{\infty} P(error|x)p(x)dx \tag{8.4}$$

and if for every x we ensure that $P(error|\underline{x})$ is as small as possible, then the integral must be as small as possible. Thus we have justified the following *Bayes decision rule* for minimizing the probability of error:

$$\text{Decide } \omega_1 \text{ if } P(\omega_1) > P(\omega_2)\text{; otherwise decide } \omega_2 \tag{8.5}$$

And under this rule, Equation (8.3) becomes [1]

$$P(error|x) = \min[P(\omega_1|x), P(\omega_2|x)] \tag{8.6}$$

8.3 Simplified Form for a Normal Distribution

The problem of classification comes with the optimization problem. We want to have a threshold or boundary conditions in the space of feature vectors in order to discriminate among different classes. This boundary condition is called a decision surface (Figure 8.4). Different decision surfaces have different properties. If the decided class is ω_i but the true class is ω_j, then the decision is correct if $i = j$ and in error if not. If errors are to be avoided, it is natural to seek a decision rule that minimizes the average probability of error, that is, the *error rate*. In the previous section, we proved that if we use $g_i(\underline{x}) = P(\omega_i|\underline{x})$ in Equation (8.2) as the decision surface, then we will minimize the probability of error by using the following decision rule:

$$\text{Decide } \omega_j \text{ } if \text{ } j = \left\{ i \middle| \underset{i}{Max} \quad g_i(\underline{x}) \right\} \tag{8.7}$$

This is called Bayes minimum error classifier. The structure of a Bayes classifier is determined primarily by the conditional density function $p(\underline{x}|\omega_i)$. Of the various density functions that have been investigated, none has received more attention than the multivariate normal density function. The general multivariate normal density is completely specified by two parameters, *mean* μ vector and *covariance* Σ matrix:

$$f(\underline{x}) = \frac{1}{(2\pi)^{d/2}|\Sigma|^{1/2}} \exp\left[-\frac{1}{2}(\underline{x} - \underline{\mu})^t \Sigma^{-1} (\underline{x} - \underline{\mu}) \right] \tag{8.8}$$

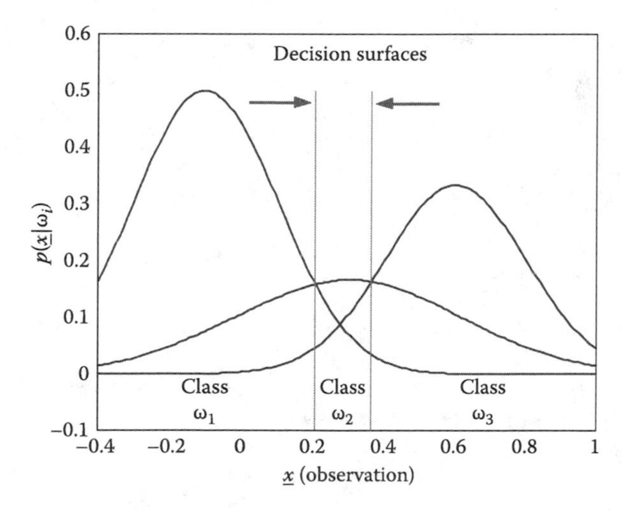

FIGURE 8.4
Decision surfaces and normal distributions.

where

$$\begin{cases} \underline{\mu} = E[\underline{x}] \\ \Sigma = E\left[\left(\underline{x} - \underline{\mu}\right)\left(\underline{x} - \underline{\mu}\right)^t\right] \end{cases} \tag{8.9}$$

For simplicity it is often abbreviated as $f(\underline{x}) \sim N(\mu,\Sigma)$. It is shown that by using normal distribution in our analysis, we can have a very simple form for the decision surface function of the Bayes minimum error classifier. This simple form is given next:

$$g_i(\underline{x}) = -0.5\left(\log|\Sigma_i| + \left(\underline{x} - \underline{\mu}_i\right)^t \Sigma_i^{-1}\left(\underline{x} - \underline{\mu}_i\right)\right) \tag{8.10}$$

where x is the feature vector, and Σ and μ are the covariance matrix and the mean vector for each class.

8.4 Feature Extraction for Our Fault Diagnosis System

Based on the discussion and analysis of faults in the previous chapters, we can decide how to choose a valid feature that carries enough information to be classified. For an induction machine, eccentricity fault affects fundamental and slot harmonics in a nonlinear form and generates some extra harmonics [2]. Also, broken rotor bar fault has an effect on the fundamental and generates side-band harmonic around the fundamental. Therefore, it makes sense if we look for features around these harmonics. In our approach, we used both the amplitude and phase of the harmonics of interest. Figure 8.5

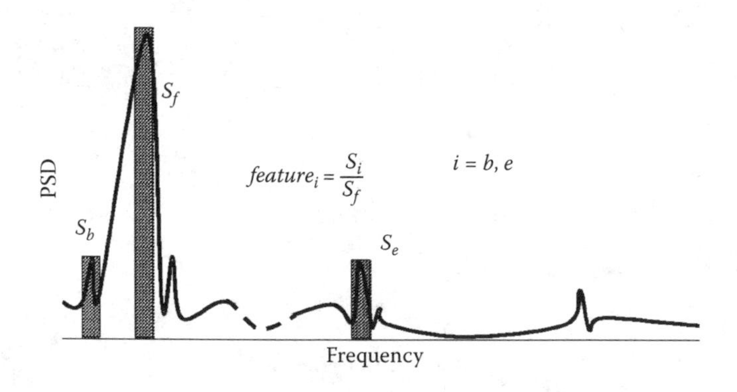

FIGURE 8.5
Extracting features from the PSD of line currents.

shows how broken rotor bar and eccentricity harmonics are used to generate two of the features of our feature vector. The other two are the phase information of these harmonics. Figure 8.6 shows the variation of these features for a data pool of 97 samples taken from the broken rotor bar fault, eccentricity fault, broken rotor bar and eccentricity, and healthy. Therefore, in our

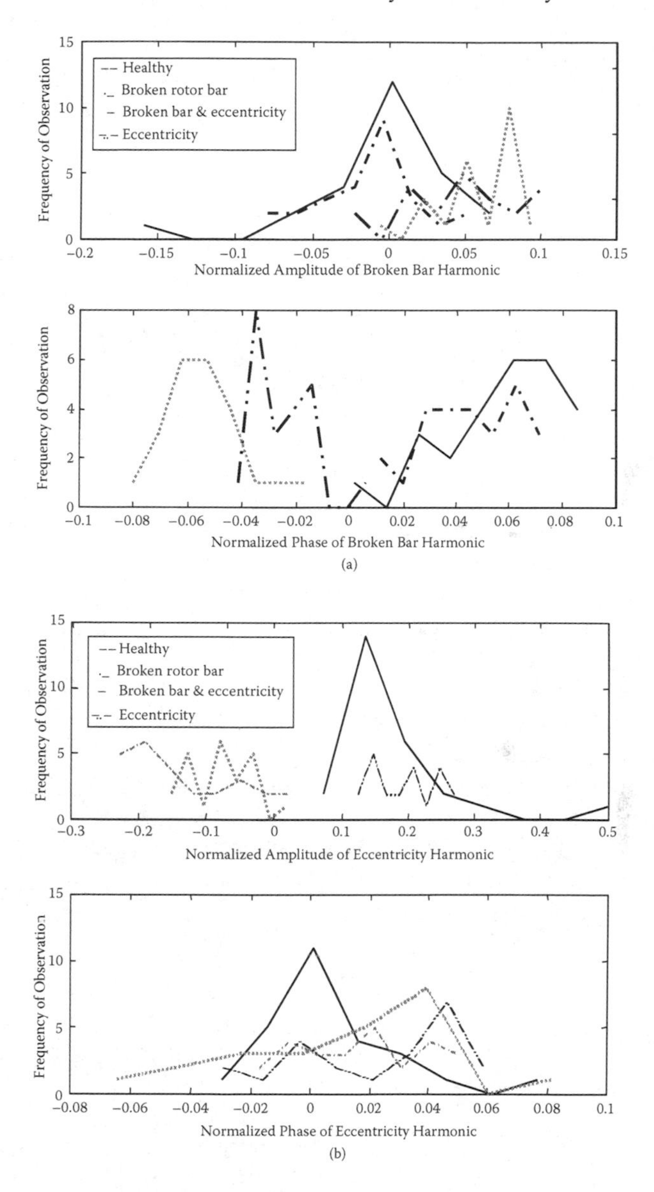

FIGURE 8.6
Variations of normalized features for a data pool of 97 samples for different conditions of an induction machine (a) broken rotor bar, (b) eccentric rotor.

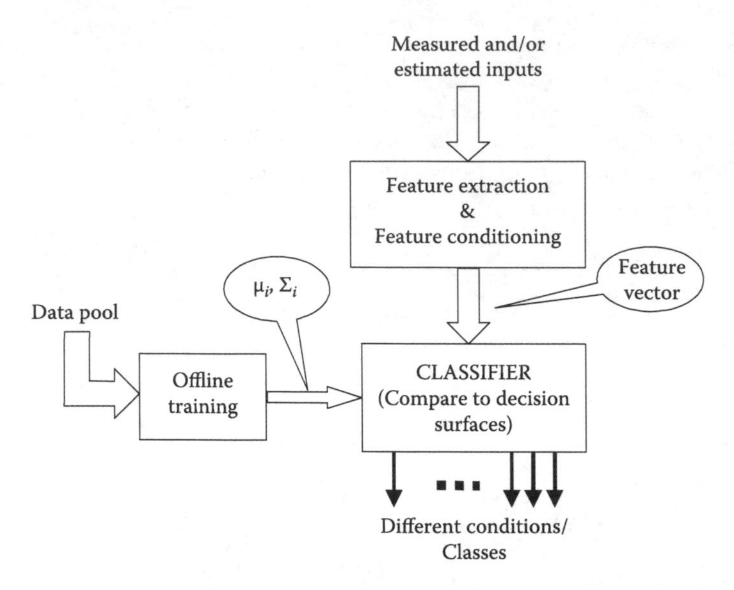

FIGURE 8.7
Block diagram of the fault diagnosis system.

experiments we had four different classes with 97 sets of data collected and processed to cover most torque-speed conditions ranging from 10% to 130% of rated load in all four cases. The utilized decision surface function has the form of Equation (8.10), where x is the feature vector and Σ and μ are the covariance matrix and the mean vector for each class (i = broken rotor bar, eccentricity, broken rotor bar and eccentricity, healthy). Figure 8.7 shows the block diagram of presented technique. Since this approach is based on a supervised learning method, we need to train the classifier before using it as an on-line fault diagnosis package. Training means finding the values of Σ_ι and μ_ι for each class and is discussed in the next section.

It should be highlighted that extracting both broken rotor bar and eccentricity harmonics needs the rotor speed information. Either a speed sensor or sensorless methods can be used for detection or estimation of the rotor speed.

8.5 Classifier Training

As is shown in Figure 8.7, features are extracted from the measured or estimated inputs. In our fault diagnosis system of induction machines, stator current is measured first and then its power spectral density (PSD) is calculated. Features are extracted from PSD by the way explained in the previous section. After being extracted, features are normalized and fed to our trained classifier. Applying the discriminant function or decision surface of Equation (8.10) on the produced feature vector, this classifier decides whether data comes from a healthy motor or a faulty one (broken rotor bar, eccentricity, or broken rotor bar and eccentricity).

Normalization of the features is important in order to exclude effects of load torque from the features and to make a numerical meaningful feature vector, that is, having values from –1 to 1. This normalization is done in the following two steps:

1. First, the integral of PSD in a window around harmonics of interest and fundamental harmonics is calculated. Then, division of these two gives a normalized power of the feature (Figure 8.5).
2. In order to give the same weight to different features, all produced data from healthy and all faults are put together and normalized according to Equation (8.11).

$$\begin{cases} \hat{\mu}_j = \dfrac{1}{N_h + N_b + N_{ec} + N_{be}} \displaystyle\sum_{i=1}^{N_h+N_b+N_{ec}+N_{be}} \hat{x}_{ij} \\ \hat{\sigma}_j = \dfrac{1}{N_h + N_b + N_{ec} + N_{be} - 1} \displaystyle\sum_{i=1}^{N_h+N_b+N_{ec}+N_{be}} (\hat{x}_{ij} - \hat{\mu}_j)^2 \end{cases} \tag{8.11}$$

where $\hat{\underline{x}} = \{x_{1h}, \cdots, x_{N_h h}, x_{1b}, \cdots, x_{N_b b}, x_{1ec}, \cdots, x_{N_{ec} ec}, x_{1be}, \cdots, x_{N_{be} be}\}$ and –1 in the formula of $\hat{\sigma}_j$ is for the unbiased estimation. j is the number of features and i refers to each sample. It is assumed that all features have the same importance in this analysis, and that is why they are given the same weight in the normalization process. Motor manufacturers can do a thorough statistical investigation to see which feature has more importance. Then, they can come up with different weighting factors for the normalization of the features. Since we could not provide a very huge data pool, such investigation would have less credibility at this time. Finally, the normalized feature vector x is given by

$$x_j = \frac{\hat{x}_j - \hat{\mu}_j}{\hat{\sigma}_j}, \quad j = 1, 2, \ldots, m \tag{8.12}$$

where m is the dimension of our feature space and the feature vector is $\underline{x} = (x_1, \ldots, x_m)$. According to Bayes minimum error classifier, mean vector (μ_i) and covariance matrix (Σ_i) of each class should be known or estimated. This is called off-line training. For simplicity, we assumed that all features have normal distribution, which is a fair assumption. Therefore, the maximum likelihood estimate for the mean and covariance of each class are

$$\underline{\mu}_j = \frac{1}{N_j}\sum_{i=1}^{N_j} \underline{x}_{ij}, \quad j = h, b, ec, be \tag{8.13}$$

$$\Sigma_j = \frac{1}{N_j}\sum_{i=1}^{N_j} (\underline{x}_{ij} - \underline{\mu}_j)(\underline{x}_{ij} - \underline{\mu}_j)^t \tag{8.14}$$

There should not be any confusion between $\hat{\mu}_j$ and $\hat{\sigma}_j$ in Equation (8.11) with μ_i and Σ_i in Equation (8.13) and Equation (8.14). The first two are used for normalizing the extracted feature vector ($\underline{x}$), whereas the last two are used in the decision surface function for the classification in Equation (8.10).

8.6 Implementation

This technique was implemented for the fault diagnosis of an induction machine shown in Figure 8.10. To have a faulty machine, the bearing housing of the end shield was machined off center, as is shown in Figure 8.8. This generates a static eccentricity. To generate a broken rotor bar fault, the rotor bars were simply broken by a drill (Figure 8.9).

FIGURE 8.8
Generating eccentricity fault by off-center machining the bearing house of the end shield.

FIGURE 8.9
Broken rotor bar generation in the rotor of an induction machine.

As mentioned before, we could collect 97 different sets of data taken from 10% to 130% of rated load in all four cases (i.e., healthy, broken rotor bar, eccentricity, and mixed broken rotor bar and eccentricity). The motor was a 3 hp, 3-phase induction motor with a rotor having 44 bars. Since there were not very many samples for training and testing the classifier, one sample was taken off each time and the classifier was trained with the other 96 samples. Table 8.1 shows the diagnostic results having a Bayes minimum error classifier and a feature vector as mentioned before. Note that three samples are misclassified between the broken and broken and eccentricity classes.

FIGURE 8.10
The test bed used for implementing our fault diagnosis system at the Electrical Machines & Power Electronics (EMPE) lab at Texas A&M University.

TABLE 8.1
Classification Results

Classified Samples →	Healthy	Broken Rotor Bar Fault	Static Eccentricity Fault	Broken Rotor Bar and Static Eccentricity Fault
Real Samples↓				
Healthy (23)	21	2	0	0
Broken Rotor Bar Fault (26)	0	24	0	2
Static Eccentricity Fault (22)	0	2	20	0
Broken Rotor Bar and Static Eccentricity Fault (26)	0	1	0	25

Note: The misclassification error is 7.2%. Note that three samples were misclassified between the broken rotor bar, and broken rotor bar and eccentricity classes.

This is compatible with what Filippetti et al. [3] proved, that broken rotor bar harmonics also carry eccentricity information.

The technique of Figure 8.7 can be applied to the fault diagnosis of a DC motor. In a DC motor, eccentricity fault affects amplitude and phase of some harmonics in the line current signal [4]. In fact, it generates several harmonics as well as affecting those slot harmonics in the back-electromotive force (EMF) signal:

$$f_{sh \quad in \ DC \ Machine} = \frac{\omega \times kR}{2} \tag{8.15}$$

Features are extracted from the PSD of the armature current signal similar to the way explained in Figure 8.5, except that it is only applied to the slot harmonic and DC, which in here is the fundamental. The experiment is performed on a 3 hp–shunt DC motor with a rotor having 20 slots. The DC motor was operated at different speeds under healthy and dynamic eccentricity conditions. In order to generate an eccentricity fault, a heavy weight was attached to one side of a disk coupled to the shaft. All features are normalized to exclude the effect of load so there can be comparable data for different conditions (Equation 8.16 and Equation 8.17).

$$\begin{cases} \hat{\mu}_j = \dfrac{1}{N_h + N_{ec}} \displaystyle\sum_{i=1}^{N_h+N_{ec}} \hat{x}_{ij} \\ \hat{\sigma}_j = \dfrac{1}{N_h + N_{ec} - 1} \displaystyle\sum_{i=1}^{N_h+N_{ec}} (\hat{x}_{ij} - \hat{\mu}_j)^2 \end{cases} \tag{8.16}$$

where $\underline{\hat{x}} = \{x_{1h}, \cdots, x_{N_h h}, x_{1ec}, \cdots, x_{N_{ec} ec}\}$ and –1 in the formula of $\hat{\sigma}_j$ is for the unbiased estimation. j is the number of features and i refers to each sample. Finally, the normalized feature vector x is given by

$$x_j = \frac{\hat{x}_j - \hat{\mu}_j}{\hat{\sigma}_j}, \quad j = 1, 2, \ldots, m \tag{8.17}$$

where m is the dimension of our feature space and the feature vector is $x = (x_1, \ldots, x_m)$. The classifier is a Bayes minimum error classifier with the assumption that all features have normal distribution with equal likely classes.

Since not very much data for training and test were available, one suitable approach was to take one sample out from the data pool as a test sample and use the remaining for training the classifier. This procedure was iterated 16 times and results were compared with the true classes. Table 8.2 shows the classification results. Three samples were misclassified, which means we have nearly 18.75% error. This is because of the limited number of training and test samples. We can expect that the actual error in on-line monitoring is much less.

In this chapter, a fault diagnosis system was shaped based on the results provided from previous chapters and a well-known pattern recognition technique. Bayesian decision theory was introduced first and used as a core to detect and classify possible faults. Analytical and experimental results yield to a very good combination of features, taken from the power spectral density of the line current of both AC and DC machines. The faults being investigated and looked at were mainly eccentricity and broken rotor bar faults. However, this technique is quite useful for detecting other types of fault as long as proper features are gathered and utilized.

The technique requires a rich data pool, which for motor manufacturers is not a difficult task. Even with the limited data we could obtain in the Electrical Machines & Power Electronics (EMPE) lab at Texas A&M University, this approach proved to be feasible due to its satisfactory results.

TABLE 8.2

Results of the Bayes Minimum Error Classifier for Detection of the Eccentricity Fault in a DC Motor

Classified Samples →	**Healthy**	**Eccentricity Fault**
Real Samples↓		
Healthy	8	1
Eccentricity Fault	2	5

Note: The misclassification error is 18.75%.

References

[1] R.O. Duda and P.E. Hart, *Pattern Classification and Scene Analysis*, New York: Wiley, 1973.

[2] M. Haji and H.A. Toliyat, "Pattern recognition: A technique for induction machines rotor broken bar detection," *IEEE Transactions on Energy Conversion*, vol. 16, no. 4, pp. 312–317, December 2001.

[3] F. Filippetti, G. Franceschini, C. Tassoni, and P. Vas, "AI techniques in induction machines diagnosis including the speed ripple effect," *IEEE Transactions on Industry Applications*, vol. 34, no. 1, pp. 98–108, January/February 1998.

[4] M. Hajiaghajani, H.A. Toliyat, and I. Panahi, "Advanced fault diagnosis of DC motors," *IEEE Transactions on Energy Conversion*, vol. 19, no. 1, pp. 60–65, March 2004.

9

Implementation of Motor Current Signature Analysis Fault Diagnosis Based on Digital Signal Processors

Seungdeog Choi, Ph.D.
Toshiba International

Bilal Akin, Ph.D.
Texas Instruments, Inc.

9.1 Introduction

Within the last decade many studies have been conducted to detect electric machine faults prior to possible catastrophic failure [1–9]. One of the most popular methods for fault diagnosis is motor current signature analysis (MCSA) as it is more practical and less costly. Thanks to recent digital signal processor (DSP) technology developments, motor fault diagnosis can now be done in real-time based on the stator line current [10–17] allowing precise and low-cost motor fault detection. Beyond this, once simple and efficient fault detection algorithms are employed, it is possible to control the motor and detect the fault at very early stages simultaneously using the same DSP [12,17]. Typically, implementing a comprehensive fault diagnosis algorithm taking all the details into account like the decision-making stage is a long and complicated procedure. Therefore, in order to not violate CPU utilization and degrade motor control performance, the priorities of the DSP-based fault algorithms need to be carefully determined based on practical issues, including limited memory occupancy and computation complexity.

Among widely used traditional algorithms, spectrum analysis has been applied in fault diagnosis such as the fast Fourier transform (FFT), which is one of the most popular signal processing algorithms in motor fault detection applications. However, in real-time applications, $(N/2) \times \log(N)$ complexity of FFT-radix 2 brings an overwhelming burden to the DSP where significant amounts of data need to be processed to produce sufficiently high resolution. Many of conventional FFT type or time-frequency analysis techniques have a similar problem in a DSP implementation. Using some of the recently proposed signal processing algorithms as alternatives to traditional methods [16–18] gives good real-time performance and satisfactory results when implemented by a committed high-speed DSP. On the other hand, a specific fault signature analysis technique instead of wide spectral analysis such as a phase locking loop, matched filtering, reference frame theory, and other relevant techniques have lower computational complexity for processing a large amount of data. Cruz et al. successfully implemented a simple algorithm based on multiple reference frame theory on a DSP used for direct torque control (DTC) of an induction machine [17]. The complexity order of a basic phase locking loop function is N, which is $\log(N)/2$ times less than that of an FFT algorithm while occupying negligible memory. Instead of scanning the whole spectrum, a phase locking loop concentrates only on the expected fault frequencies that improve resolution and noise immunization [12].

9.1.1 Cross-Correlation Scheme Derived from Optimal Detector in Additive White Gaussian Noise (AWGN) Channel

While performing motor fault detection, it is important to have a noise suppression capability where high-energy noise content dominates the low amplitude fault signatures. As an effective tool, the matched filter is often pronounced as one of the best candidates [19] in an additive white Gaussian noise (AWGN) channel. The matched filter is known as an optimal detector that maximizes the signal-to-noise ratio (SNR) in the AWGN channel. A typical filter is expressed by

$$y_n = \sum_{k=1}^{N} h_{n-k} s_k \tag{9.1}$$

where $n = 1,2,\cdots N$, h_n represent the impulse response of the filter, and s_k is the input signal. The output SNR of the filter can be written as

$$SNR = (H^T S)^2 / E[(H^T W)^2] = (H^T S)^2 / (\sigma^2 H^T H) \tag{9.2}$$

where $H = [h_N, h_{N-1,\ldots} h_1]$, $S = [s_1, s_2 \cdots s_N]$, $W = [w_1, w_2, \cdots w_N]$, w_n is the sampled Gaussian noise with variance σ^2, and T is the vector transpose. Through the Cauchy–Schwarz inequality, the denominator in Equation (9.2) is maximized as

$$(H^T S)^2 \leq (H^T H)(S^T S) \quad when\ H = cS \tag{9.3}$$

where c is constant. It is obvious from Equation (9.3) that the SNR of filtering is maximized when $h_n = s_{N-n}$, which is called the matched filter.

Assuming s_n is the reference signal of the inspected fault signature and x_k is the input current signal, the output of matched filter is rewritten in the form of cross-correlation as given in Equation (9.4), which is supposed to suppress noise optimally for fault signature detection. Hence, cross-correlation can be proposed as one of the best signal detectors for the systems distorted by Gaussian noise.

$$y_n = \sum_{k=1}^{N} h_{n-k} s_k = \sum_{k=1}^{N} h_{n-k} x_k = \sum_{k=1}^{N} s_{N-(n-k)} x_k \quad \Rightarrow y_N = \sum_{k=1}^{N} s_k x_k \tag{9.4}$$

The analysis of a matched filter in continuous time can be derived in a similar manner through integration instead of summation in Equations (9.1) to (9.4) [19]. The matched filter output in continuous time can also be expressed as the cross-correlation in Equation (9.4) by replacing the summation to integration. The details of continuous-time matched filters are not covered since the implementation is based on discrete time processing.

Implementation of an algorithm on DSP is commonly limited by the memory and computing capacity of the system. The memory occupancy for cross-correlation operation is assumed negligible because it is performed in the sample sequence order of the input signal x_k in Equation (9.4), which does not need an additional signal memory buffer. The computing complexity of the cross-correlation is shown as N in Equation (9.4), which is low enough as each multiplication occurs only one time in each interrupt in normal operation of a DSP system. For the FFT-based scheme, which has been popularly used in fault diagnosis, all the signals should be inherently stored in a memory buffer for computation and the number of multiplication required is $(N/2) \times \log(N)$, which is assumed not acceptable due to the overwhelming burden on DSP, especially for low cost on-line fault diagnosis systems. The inherent optimal performance in noise suppression, low memory occupation, and low computing complexity makes the cross-correlation based detection an attractive tool for on-line fault diagnosis of a motor.

Most of the specific fault signal detection schemes in literature utilize the optimal property of the matched filtering such as reference frame theory, phase sensitive detection, and any other relevant cross-correlation methods. In this chapter, the implementation of reference frame theory and phase-sensitive detection as an example in an embedded DSP system is presented.

9.2 Reference Frame Theory

The topic of phase transformations and reference frame theory [12] constitutes an essential aspect of machine analysis and control. In this chapter, apart from the conventional applications, it is reported that the reference frame theory can also be successfully applied to fault diagnosis of electric machinery systems as a powerful toolbox to find the magnitude and phase quantities of fault signatures. The basic idea is to convert the associated fault signature to direct current (DC) quantity, followed by the computation of the signal's average in the new reference frame to filter out the rest of the signal harmonics, that is, its alternating current (AC) components. Because the rotor and stator fault signature frequencies are well known, the presented method focuses only on the fault signatures in the current spectrum, depending on the examined motor fault.

9.2.1 Reference Frame Theory for Condition Monitoring

The introduction of reference frame theory in the analysis of electrical machine systems has turned out not only to be useful in their control and analysis, but also has provided a powerful tool for condition monitoring. By judiciously choosing the reference frame, it is possible to monitor any kind of motor fault whose effects are reflected to the line current as shown in the following section. The rotating reference frame module in the software used for fault analysis can work separately and independently than the one used for motor control, which is synchronized to the fundamental harmonic vector.

9.2.2 (Fault) Harmonic Analysis of Multiphase Systems

The commonly used transformation is the polyphase to orthogonal two-phase transformation. The complex current harmonic vector describes a circular trajectory in the space vector plane as shown in Figure 9.1. Therefore, a multiphase system in phase variables transforms to a circular locus in the equivalent two-axis representation. In Figure 9.1, the radius of the circle around the origin is the peak magnitude of the inspected harmonic quantities, and the vector rotational frequency is equal to the angular frequency

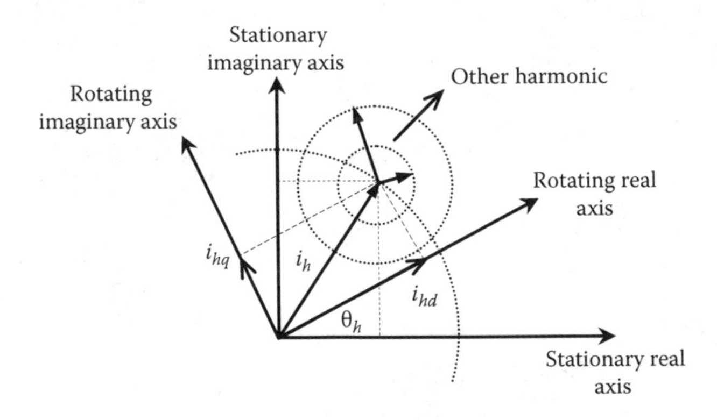

FIGURE 9.1
Harmonic space vector with other harmonic vectors in the stationary and rotating reference frames.

of phase harmonic quantities. Note that the drawings in Figure 9.1 are exaggerated to explain the basis of the theory explicitly; indeed the magnitude of fundamental harmonic is several times higher than all line and fault harmonics. If the new rotating reference frame is defined where the axes are made to rotate at the same rate as the angular frequency of the inspected harmonic, a stationary current space vector results, where its orthogonal components are DC quantities.

If a reference frame is synchronized to a particular frequency, in the new reference frame all harmonics other than the inspected one remain as AC. The average of these AC harmonics converge to zero and have a negligible effect on the average after a sufficient time. In other words, the reference frame synchronized with fault harmonic shifts the frequency spectrum of the phase current by frequency of the fault component. The rotating frame converts only the associated fault harmonic vector to a stationary vector at zero Hertz whose projection on orthogonal base vectors are DC and the averages are nonzero in time. Thus, when the resultant fault vector modulation is normalized with respect to the fundamental vector that is computed at the synchronously rotating reference frame, the ratio gives the relative magnitude of the fault harmonic as

$$\left|\frac{I_{fault}}{I_1}\right| = 20\log\left[\frac{\left.\left(\left(\sum_k i_{dk}\right)^2+\left(\sum_k i_{qk}\right)^2\right)\right|_{\theta_h=\theta_{fault}}}{\left.\left(\left(\sum_k i_{dk}\right)^2+\left(\sum_k i_{qk}\right)^2\right)\right|_{\theta_h=\theta_1}}\right]^{1/2} \quad \text{(db)} \tag{9.5}$$

where i_{dk} and i_{qk} are the dq components of phase current in the rotating frame, θ_1 is the angular position of the stator reference frame, and I_1 and I_{fault} are the relative magnitudes of the fundamental and fault harmonic vectors, respectively. The fundamental operation of detection in Equation (9.5) constitutes multiple cross-correlation operation (matched filtering). In addition to fault harmonic magnitude calculation, the phase angle information of associated harmonic vector can also be found using the direct (d) and quadrature (q) components obtained by the proposed technique. The dq components of the harmonic vectors decouple depending on the phase angle between the rotating frame and the vector as shown in Figure 9.1. Therefore, the phase angle is formulated as

$$\varphi_{fault} = \tan^{-1}\left(\frac{\left.\sum_k i_{hqk}\right|_{\theta_h=\theta_{fault}}}{\left.\sum_k i_{hdk}\right|_{\theta_h=\theta_{fault}}}\right) \tag{9.6}$$

One must note that the notations, indexes, and axes of the frames might change depending on how they are defined by the user. In the literature there are different representations of reference frame theory, but the basics are the same. So reference frame theory can be further simplified for single-phase signal-based detection, which can be derived without loss of generality.

9.2.3 On-Line Fault Detection Results

Experiments are done on-line using the TMS320F2812 DSP, which is employed both for inverter control and fault signature detection. Several experiments are realized under various conditions such as different rotor speeds, slip, load conditions, switching frequencies, sampling frequency, and the number of data processed.

When using DSP core for both control and fault purposes, the fault code is embedded into the main control algorithm as a subroutine that processes the instantaneously measured current data for both fundamental component and fault signature frequency. The same experiments are also repeated for line-driven motors where the DSP is responsible only for fault analysis rather than control issues. Although undersampling and oversampling are possible, generally switching frequency is accepted to be the sampling frequency of the current data to synchronize the fault subroutine with the main control. The number of data is chosen to be the same as the sampling frequency, which can be adjusted between 4K and 20K depending on the applications. The stator frequency can either be calculated or equated to the reference value depending on the control type, and the rotor speed can either be measured using encoder or estimated to update the signature frequencies in real time. Though the DSP of the inverter is used in this experiment, the very simple algorithm of reference frame theory can be implemented using a simpler microcontroller as well.

9.2.3.1 v/f Controlled Inverter-Fed Motor Line Current Analysis

The eccentricity and broken rotor bar tests are repeated using TMS320F2812 DSP controlled inverter where $\omega_{rref} = 0.99$ per unit. The motor is run at no load and at full load for eccentricity tests and broken rotor bar tests, respectively. As shown in Figure 9.2, both the eccentricity and broken rotor bar sidebands found by DSP microprocessor are very close to ones observed by FFT spectrum analyzer at $(f_s \pm f_r)$ and $(1 \pm 2s)f_s$, respectively. The time spent to process 5K to 20K data and detect these signatures is 1 second, which is sufficiently short for fault monitoring where there is no strict time limitation. Depending

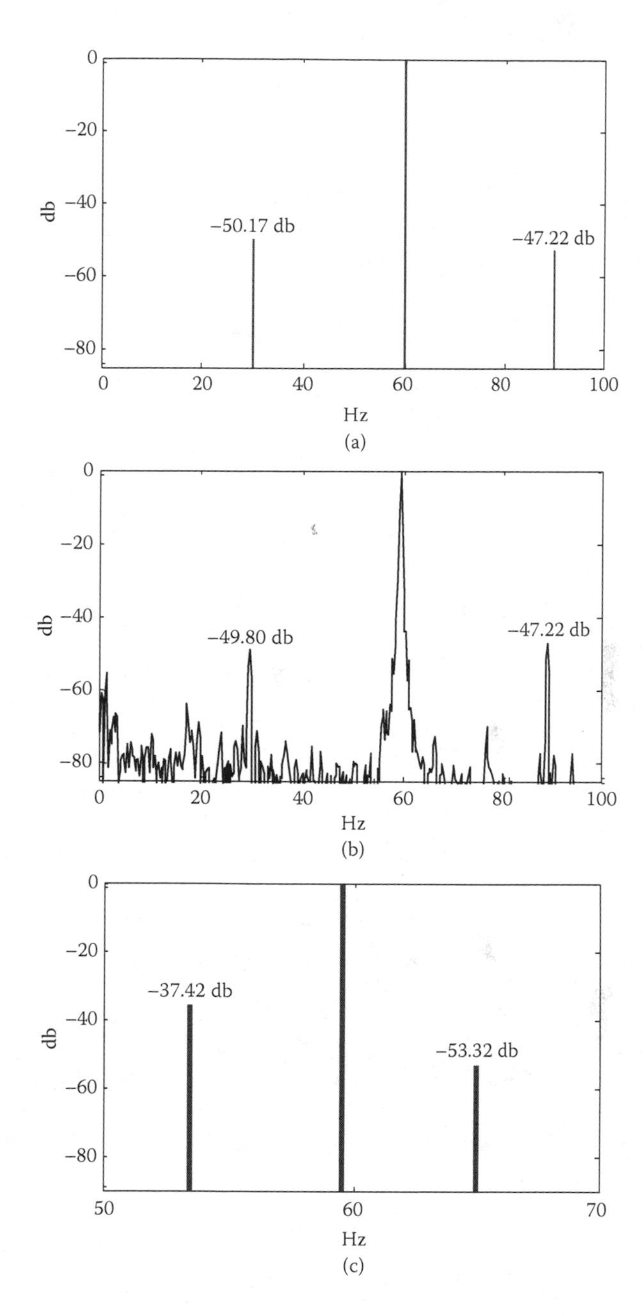

FIGURE 9.2
Experimentally obtained v/f controlled inverter-fed motor single-phase harmonic analysis result: (a) eccentricity signatures detected by DSP using rotating frame theory, (b) FFT spectrum analyzer output of eccentric motor line current, (c) broken rotor bar signatures detected by DSP using rotating frame theory, (d) FFT spectrum analyzer output of broken rotor bar motor line current.

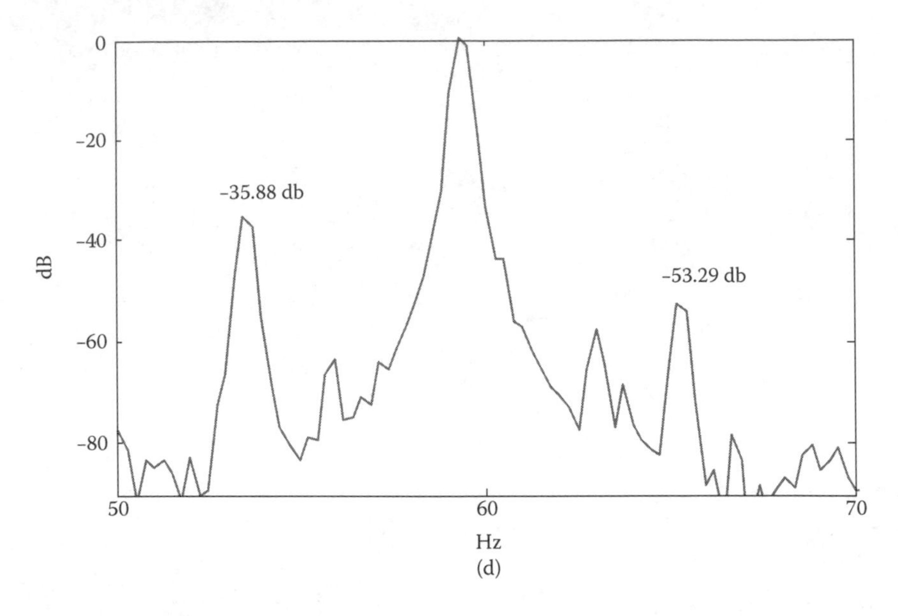

FIGURE 9.2 *(continued)*
Experimentally obtained v/f controlled inverter-fed motor single-phase harmonic analysis result: (a) eccentricity signatures detected by DSP using rotating frame theory, (b) FFT spectrum analyzer output of eccentric motor line current, (c) broken rotor bar signatures detected by DSP using rotating frame theory, (d) FFT spectrum analyzer output of broken rotor bar motor line current.

on the resolution requirements and the system control parameters, execution time might be shortened or extended.

9.2.3.2 Field-Oriented Control Inverter-Fed Motor Line Current Analysis

In Figure 9.3, the same experiments are repeated running the motor with closed-loop field-oriented control algorithm at various operating points. The results obtained by industry purpose processor and 12-bit analogue-to-digital converter (ADC) are very close to FFT spectrum analyzer outputs, which have two DSP core and 16-bit ADC with a sampling rate of 256 kHz. These on-line experimental results confirm that the presented method can be successfully adapted to the real-time applications.

9.2.3.3 Instantaneous Fault Monitoring in Time-Frequency Domain and Transient Analysis

A stationary motor line current signal repeats into infinity with the same periodicity. However, this assumption is not realistic for most of the industrial applications where the duty cycle profile of the motor cannot be guaranteed to operate at steady state and at a single operating point. Instead, duty

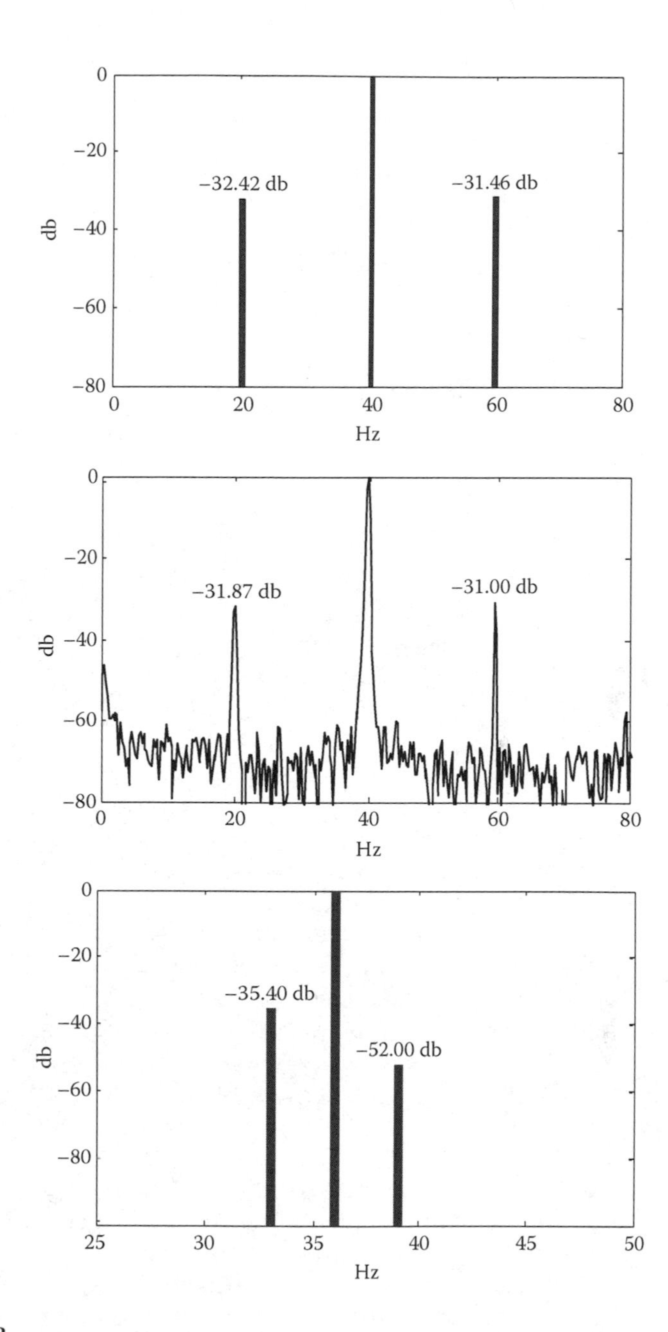

FIGURE. 9.3
Experimentally obtained FOC controlled inverter-fed motor single-phase harmonic analysis result: (a) eccentricity signatures detected by DSP using rotating frame theory, (b) FFT spectrum analyzer output of eccentric motor line current, (c) broken rotor bar signatures detected by DSP using rotating frame theory, (d) FFT spectrum analyzer output of broken rotor bar motor line current.

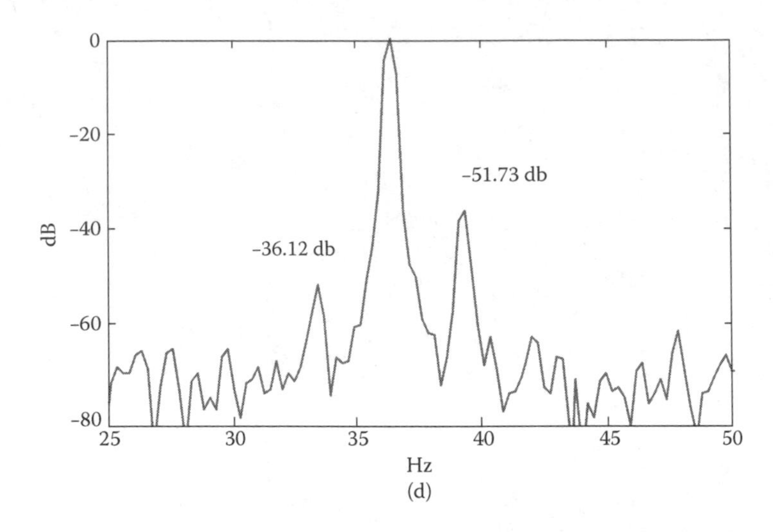

FIGURE. 9.3 (*continued*)
Experimentally obtained FOC controlled inverter-fed motor single-phase harmonic analysis result: (a) eccentricity signatures detected by DSP using rotating frame theory, (b) FFT spectrum analyzer output of eccentric motor line current, (c) broken rotor bar signatures detected by DSP using rotating frame theory, (d) FFT spectrum analyzer output of broken rotor bar motor line current.

cycle involves various operating points at different load and speed combinations for an unknown time period.

On the other hand, the motor current spectrum analyses done using Fourier transform assumes that the current signal is stationary. The Fourier transform performs poorly when this is not the case. Furthermore, the Fourier transform gives the frequency information of the signal, but it does not tell us when in time these frequency components exist. The information provided by the integral corresponds to all time instances because the integration is done for all time intervals. It means that no matter where in time the frequency appears, it will affect the result of the integration equally. This is why traditional application of Fourier transform is not suitable for nonstationary signals.

As stated earlier, continuous stator frequency and shaft speed information are available and are used to update fault signature frequencies at all operating points. The updated fault signature frequency is utilized to synchronize the reference frame and associated fault vector component of the line current. Therefore, even though the motor supply frequency or rotor shaft speed change due to acceleration, deceleration, loading, and so forth, the normalized fault signature magnitude is instantaneously and continuously monitored without using additional algorithms. In brief, this advantage provides real-time tracing of fault signature components in the frequency domain. In Figure 9.4a,b, the right eccentricity sideband magnitude and rotor speed are

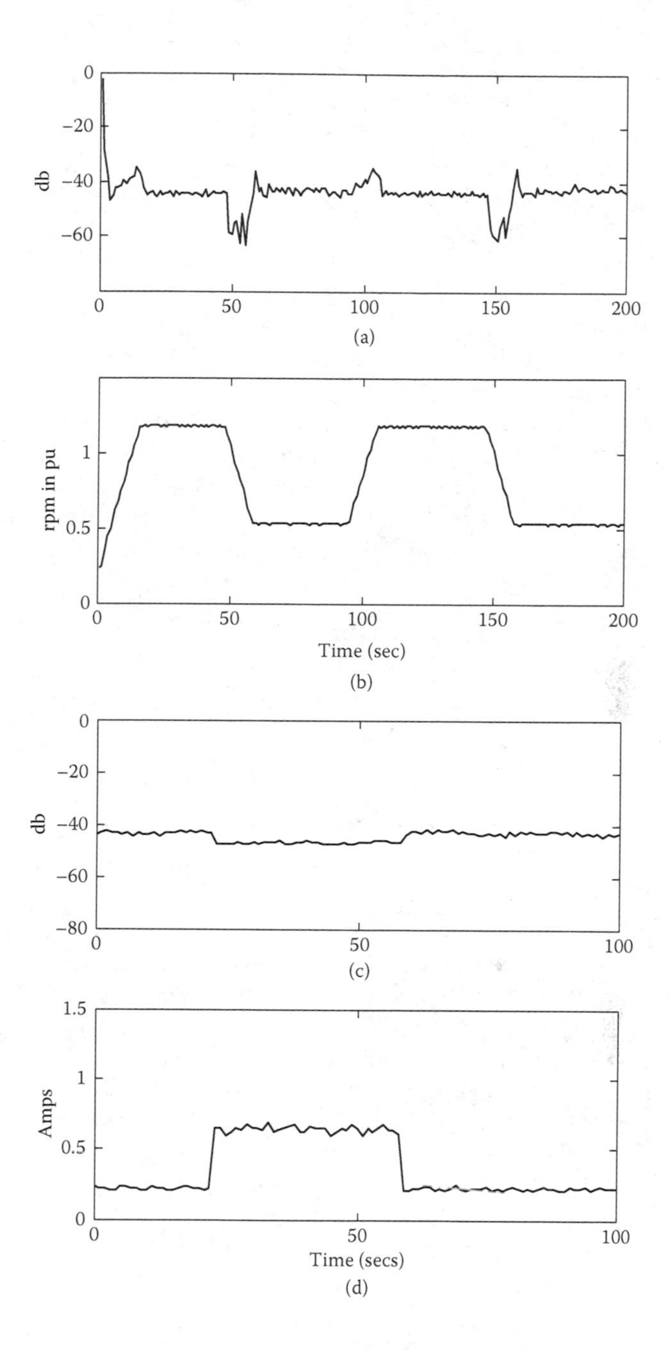

FIGURE 9.4
Experimentally obtained v/f controlled inverter-fed motor single-phase harmonic analysis result: (a) normalized eccentricity sideband variation detected by DSP using rotating frame theory, (b) motor speed in pu, (c) normalized eccentricity sideband variation detected by DSP using rotating frame theory, (d) motor line current in amps.

shown, respectively. The dynamic characteristics of the right eccentricity sideband at transients and different rotor speeds are traced experimentally by the DSP in real-time as shown in Figure 9.4a. A similar test is done under load when the motor is driven by the voltage-to-frequency (v/f) open loop control at 0.4 pu speed as shown in Figure 9.4c,d. In Figure 9.4c, the eccentricity right sideband track in real time is shown when the motor is loaded while running at no-load, and in Figure 9.4d the phase-A current vector magnitude is shown to identify the load characteristics.

This chapter has presented the experimental and the analytical validation of the reference frame theory application to electric motor fault diagnosis. The presented method has many advantages over existing fault diagnosis methods using external hardware and powerful software tools. The experimental test results are compared with FFT spectrum analyzer results to confirm the accuracy of this method. It is experimentally shown that this simple fault diagnosis algorithm can be embedded in the main control subroutine and run by the motor drive processor in real-time without affecting control performance of the inverter. Therefore, it can be considered as a no-cost application, which is highly promising for fault diagnosis products.

9.3 Phase-Sensitive Detection-Based Fault Diagnosis

9.3.1 Introduction

This section presents DSP-based phase-sensitive motor fault signature detection [21]. The implemented method has a powerful line current noise suppression capability while detecting the fault signatures. Because the line current of inverter-fed motors involves low order harmonics, high frequency switching disturbances, and the noise generated by harsh industrial environment, the real-time fault analyses yield erroneous or fluctuating fault signatures. This situation becomes a significant problem when SNR of the fault signature is quite low. It is theoretically and experimentally shown that the method can determine the normalized magnitude and phase information of the fault signatures even in the presence of noise, where the noise amplitude is several times higher than the signal itself.

9.3.2 Phase-Sensitive Detection

Phase-sensitive detection is based on correlation of two signals. In the correlation process, the input signal is compared with a reference signal and similarity between these signals is determined. Similarly, a lock-in detector takes a periodic reference signal and a noisy input signal, and then extracts only that part of the output signal whose frequency and phase match the reference. To see how the phase-sensitive detector works, consider a reference signal, I_{ref}, which is a pure sine wave with frequency of w_{ref},

$$I_{ref}(t) = I_{ref}\cos(w_{ref}t + \phi_{ref}) \tag{9.7}$$

and the noisy fault signal,

$$I_{in}(t) = I_{fault}\cos(w_{fault}t + \varphi_{fault}) + \sum I_{noise}\cos(w_{noise}t + \varphi_{noise}) \tag{9.8}$$

The correlation between these two signals is given by

$$\begin{aligned} I_{II}(\varphi) &= I_{ref}\cos(w_{ref}t + \varphi_{ref})I_{fault}\cos(w_{fault}t + \varphi_{fault}) \\ &+ I_{ref}\cos(w_{ref}t + \varphi_{ref})\sum I_{noise}\cos(w_{noise}t + \varphi_{noise}) \\ &= I_{ref}I_{fault}\cos(w_{ref}t - w_{fault}t + \varphi_{ref} - \varphi_{fault}) \\ &+ I_{ref}I_{fault}(w_{ref}t + w_{fault}t + \varphi_{ref} + \varphi_{fault}) \\ &+ I_{ref}I_{noise}\sum\cos(w_{ref}t - w_{noise}t + \varphi_{ref} - \varphi_{noise}) \\ &+_{ref} I_{noise}\sum\cos(w_{ref}t + w_{noise}t + \varphi_{ref} + \varphi_{noise}) \end{aligned} \tag{9.9}$$

The generated reference signal frequency is set to be the same as the fault signal frequency; therefore some of the terms in Equation (9.9) are converted to DC as given by Equation (9.10):

$$\begin{aligned} I_{II}(\varphi) &= I_{ref}I_{fault}\cos(\varphi_{ref} - \varphi_{fault}) + I_{ref}I_{fault}(2w_{ref}t + \varphi_{ref} + \varphi_{fault}) \\ &+ I_{ref}I_{noise}\sum\cos(w_{ref}t - w_{noise}t + \varphi_{ref} - \varphi_{noise}) \\ &+ I_{ref}I_{noise}\sum\cos(w_{ref}t + w_{noise}t + \varphi_{ref} + \varphi_{noise}) \end{aligned} \tag{9.10}$$

If the correlation output is low pass filtered simply by averaging, only two terms survive: the DC term due to the output of the system and the noise component with frequency near the reference signal. The rest of the noise and low order harmonics disappear as shown in Equation (9.11):

$$I_{II\_filtered}(\varphi) \approx K_1\cos(\varphi_{ref} - \varphi_{fault}) + K_2\sum\cos(\varphi_{ref} - \varphi_{noise}) \tag{9.11}$$

The phase of the noise signal varies randomly. In order to minimize the effects of noise content at the same frequency, the phase angle difference between the reference signal and the fault signals should be minimized. There are some alternatives to maximize the low pass filtered portion of the autocorrelation function. One alternative is to track the autocorrelation function and detect the peak point where the phase angles of the reference signal and the fault signal are the same. The second and more efficient method is examining both the correlation of cosinusoidal and sinusoidal reference signals to the same phase angle instantaneously. The arctangent of the correlation ratio results in the phase angle difference between the reference signal and the fault signal. The maximum correlation degree and minimum noise effect are observed when the phase angles are equated to each other by simply adjusting the reference signal's phase angle. The similar processes are repeated for the fundamental component to calculate the correlation ratio between the fundamental and fault components to find the normalized magnitude of the fault signature.

The characteristic frequencies of the well-known motor faults are given in the literature [22–25]. The most commonly reported faults in electric machines are bearing faults, eccentricity, broken rotor bar, and stator faults. All of these faults are modeled as functions of both stator frequency and rotor speed. These two variables are mostly observed by drive systems to control the motor effectively. Therefore, the reference signals are generated according to the fault equations using the rotor speed and the supply frequency to precisely capture the associated fault signatures.

9.3.3 On-Line Experimental Results

Tests are repeated on-line using the TMS320F2812 DSP, which is employed both for inverter control and fault signature detection. When using DSP core for both control and fault purposes, the fault code is embedded into the main control algorithm as a subroutine that processes the instantaneously measured current data. The number of data is chosen to be the same as the sampling frequency, which can be adjusted between 4k and 20k depending on the applications. The stator frequency is equated to the reference value depending on the control type. The rotor speed can either be measured using the encoder or estimated to update the signature frequencies in real time. Because the embedded ADC in TMS320F2812 has 12-bit, the quantization constraints prevent sensing signals less than −65 dB. The experiments are carried out by testing broken rotor bar and eccentric motors.

The results obtained in Figure 9.5 using the DSP with 12-bit ADCs are very close to results obtained from the FFT spectrum analyzer that has a two-DSP core and a 16-bit ADC with a sampling rate of 256 kHz. The left sideband signature of an eccentric motor is measured to be −39.24 dB and −38.98 dB using the FFT analyzer and the DSP, respectively. It is reported that the fault signature magnitude is not strongly affected by the switching frequency of the inverter. Since this measurement is taken when the motor is running at the steady state, the ratio of the number of data to the switching frequency is mostly taken as unity, which provides sufficient resolution. The correlation of the fault component and the

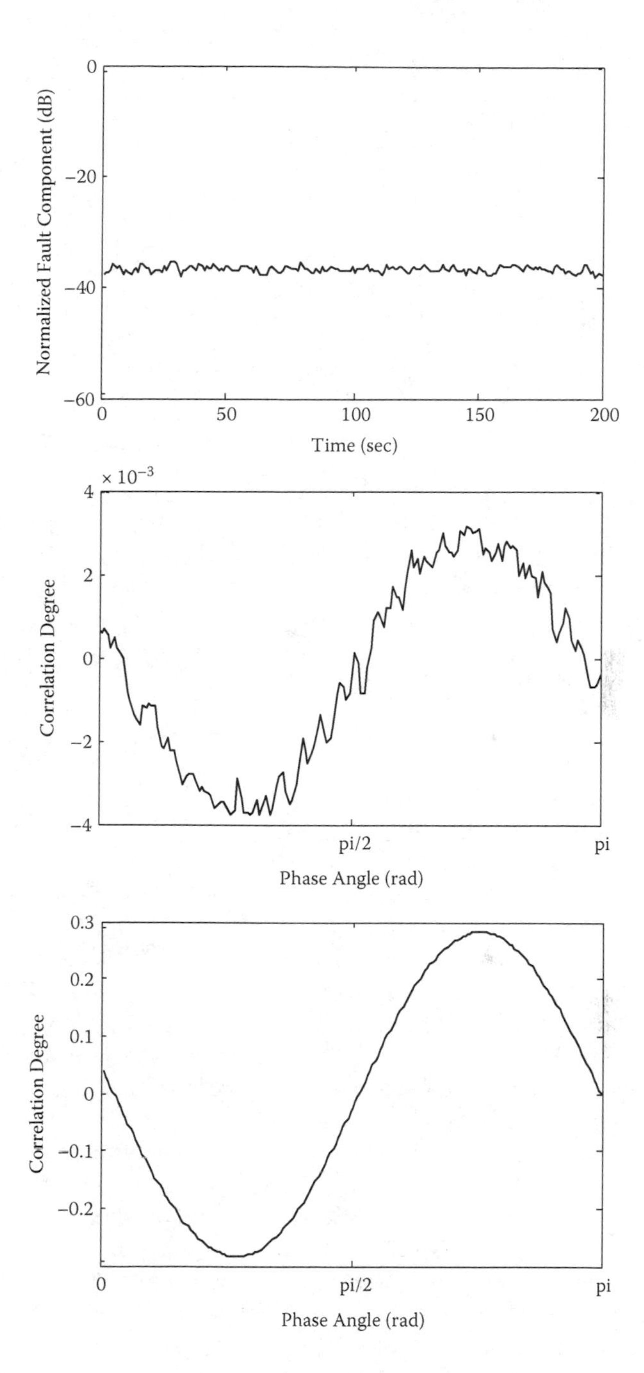

FIGURE 9.5
Experimentally obtained (a) left eccentricity sideband in real time, (b) correlation degree between reference signal and the fault component, (c) correlation degree between reference signal and the fundamental component, and (d) FFT spectrum.

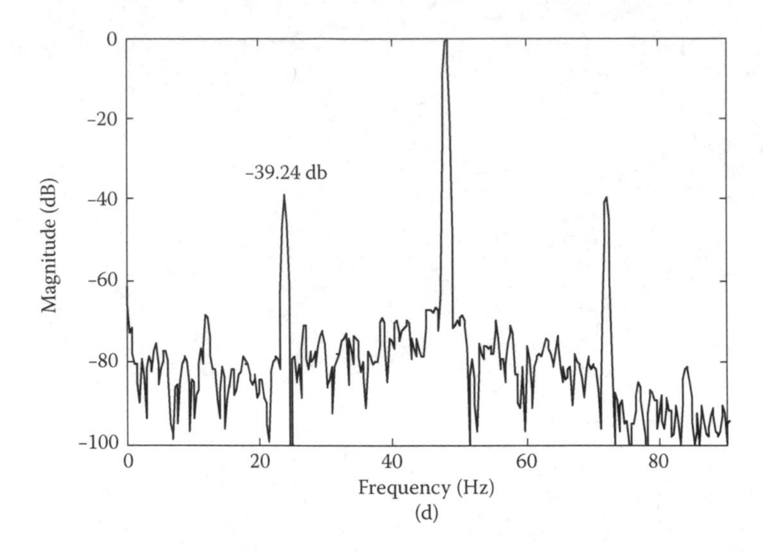

FIGURE 9.5 ***(continued)***
Experimentally obtained (a) left eccentricity sideband in real time, (b) correlation degree between reference signal and the fault component, (c) correlation degree between reference signal and the fundamental component, and (d) FFT spectrum.

fundamental component with respect to the reference signals generated by the DSP are given in Figure 9.5c,d. It is possible to obtain smoother waveforms simply by processing more data. These on-line experimental results confirm that the method can be adapted to the real-time applications.

In order to realize on-line lock-in of the reference signal and fault signature, a few ways are possible. For instance, the phase angle difference between the reference signal and the fault signature can be calculated using the arctangent relation of the cross-correlation and the autocorrelation at each fault signature detection cycle. Next, the minimum phase angle difference point is chosen as the operating point that maximizes the correlation and minimizes the noise effects. Once this point is detected, the rest of the fault diagnosis process can be continued at this point or it can be updated at each phase difference zero crossings.

As shown in Figure 9.6a, the correlation degree of fault component is set to maximum at zero crossing of the phase difference and fixed at this point until the next zero crossing. A similar process is repeated for the fundamental component to normalize the fault component as shown in Figure 9.6b. Despite the decrease in precision, the phase angle scanning can be accelerated by increasing the reference signal phase angle increments in each drive control cycle. Since the period of phase angle scanning is in the range of minutes this method is appropriate for constant duty cycle steady-state operations.

In order to examine the motors, the duty cycles of which are continuously fluctuating, an alternative autotuning algorithm is developed. Apart from the previous method, the phase difference between the reference signal and the fault signature is continuously updated. Thanks to this method, it is possible

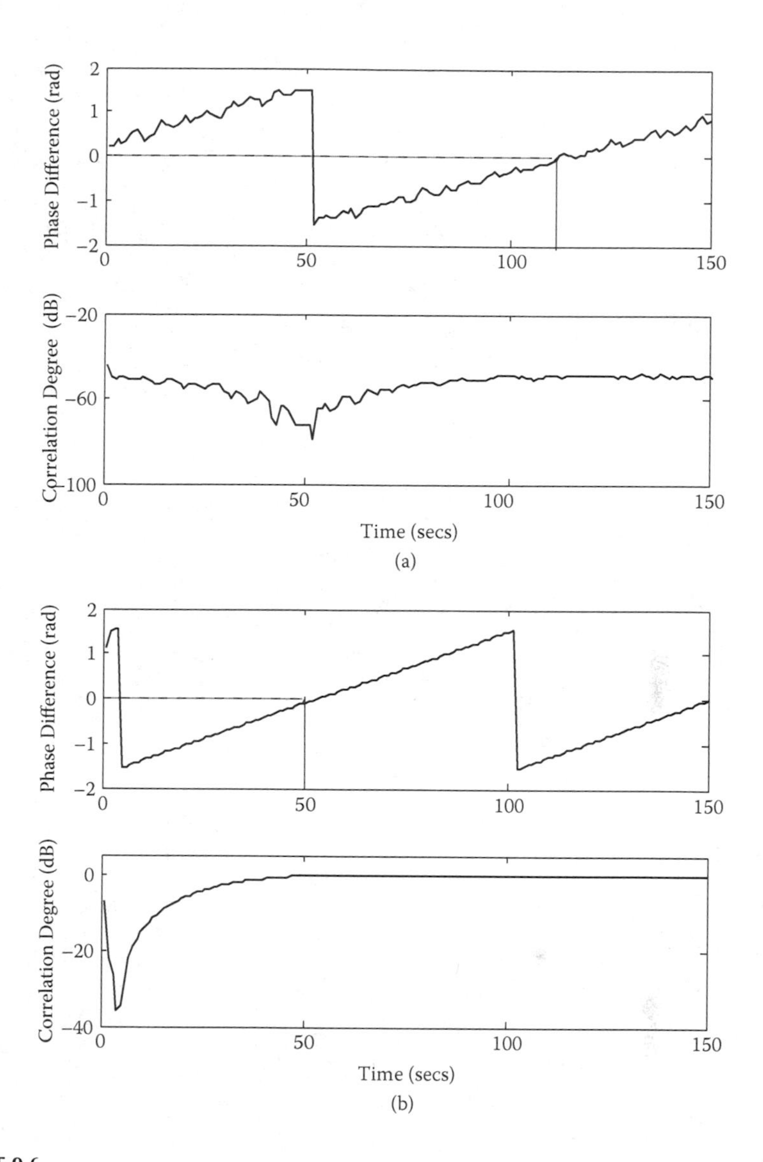

FIGURE 9.6
Experimentally obtained (a) phase difference between reference signal and fault component, and normalized left eccentricity sideband correlation degree in real time, (b) phase difference between reference signal and fault component, and normalized left eccentricity sideband correlation degree in real time.

to track fault signature not only at the steady state but during transients as well. Therefore, one can follow the dynamic characteristics of fault signatures during acceleration, deceleration, and loadings. The false error warnings can be minimized employing this method and previously determined operating point dependent on the adaptive threshold. If there are rare measurements to be made of a current magnitude that does not change significantly in time, it may be acceptable to process as many data as possible to enhance the precision of the result. However, if there are multiple measurements to be made particularly during transients, the number of processed data should be optimized. Typically, a few drive control cycles data processing time is enough at steady state and at most one or a half cycle will be sufficient during transients. It is reported that less than half a control cycle significantly degrades precision. Since the results are normalized, computation time will not affect the relative amplitude of the fault signature or the correlation degree.

In Figure 9.7, continuous tracking of the right eccentricity sideband is given. The phase lock-in is achieved in each drive control cycle by the autotuning

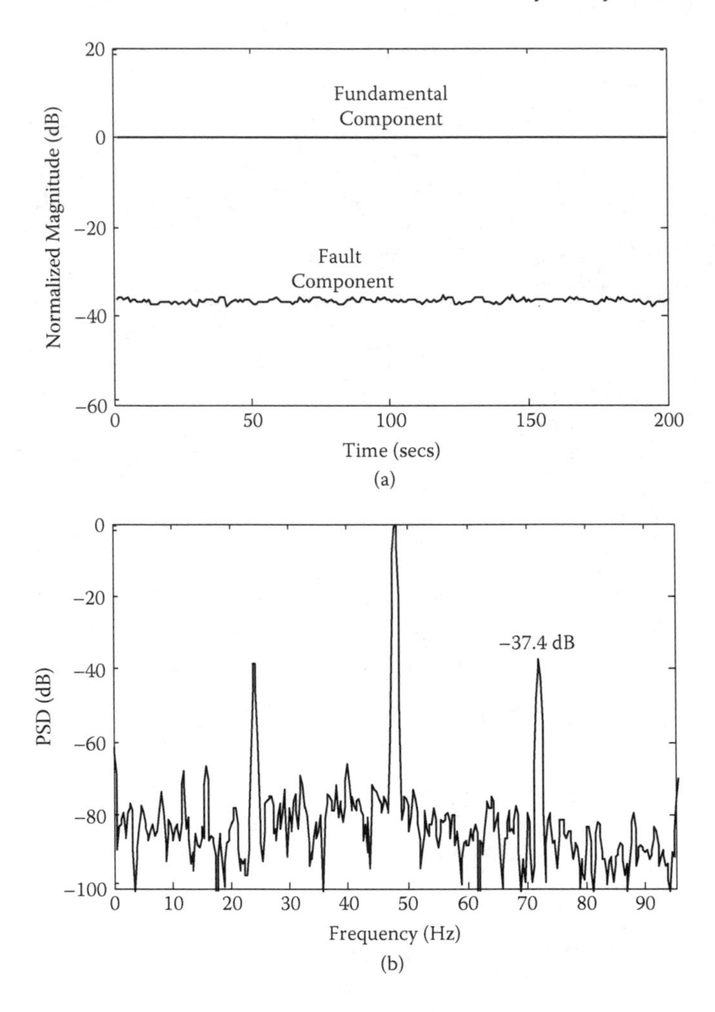

FIGURE 9.7
Experimentally obtained (a) normalized fundamental and right eccentricity sideband correlation degree in real time, and (b) FFT analyzer output.

algorithm. Using the phase-sensitive detection, the right eccentricity sideband is measured as less than |1| dB error when compared to the FFT analyzer results.

In Figure 9.8, the real-time fault signature tracks are given. In Figure 9.8a, the right eccentricity sideband variation is given from no-load to 0.33 pu

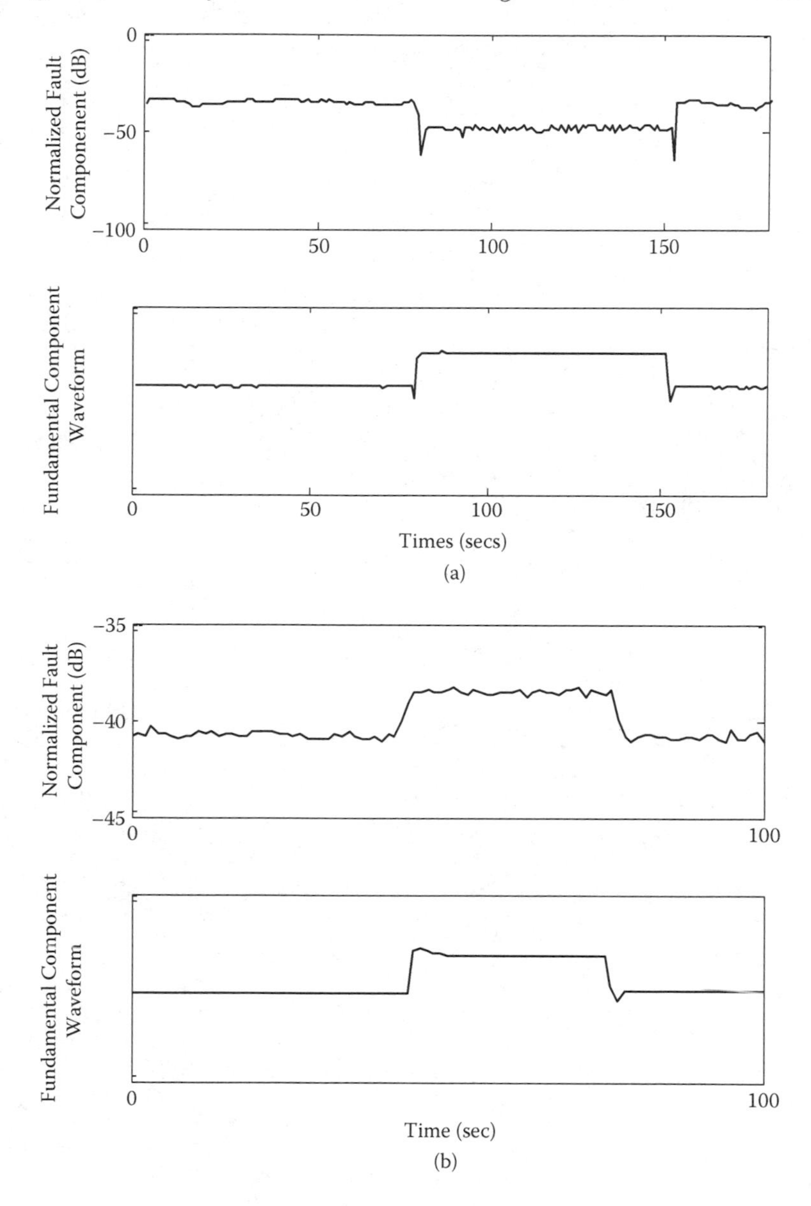

FIGURE 9.8
Experimentally obtained (a) normalized right eccentricity sideband correlation degree in real time under no load and 0.33 pu load, and (b) normalized broken rotor bar fault right sideband correlation degree in real time under 0.8 pu and 1.1 pu load.

load using the autotuned phase-sensitive lock-in detector. In Figure 9.8b, the broken rotor bar fault right sideband variation is given from 0.8 pu to 1.1 pu load. Because the supply frequency is already continuously available in the control algorithm and the rotor speed is measured or estimated, these parameters are used to update the fault signature frequencies in real time at various operating points. These results prove that the method has a powerful real-time fault signature tracking capability.

In this chapter, a simple noise immune real-time fault signature detection tool is presented. Since this method can easily be implemented using general-purpose microcontrollers without any additional hardware, PC, filters, and large size memory, it can be adapted to single- and multiphase drive systems.

References

[1] S. Choi, B. Akin, M.M. Rahimian, and H.A. Toliyat, "Fault diagnosis implementation of induction machine based on advanced digital signal processing techniques," *IEEE Transactions on Industrial Electronics*, vol. 58, pp. 937–948, March 2011.

[2] S. Nandi, H.A. Toliyat, and X. Li, "Condition monitoring and fault diagnosis of electrical machines—A review," *IEEE Transactions on Energy Conversion*, vol. 20, no. 4, pp. 719–729, December 2005.

[3] A. Siddique, G.S. Yadava, and B. Singh, "A review of stator fault monitoring techniques of induction motors," *IEEE Transactions on Energy Conversion*, vol. 20, pp. 106–114, March 2005.

[4] A. Bellini, F. Filippetti, C. Tassoni, and G.A. Capolino, "Advances in diagnostic techniques for induction machines," *IEEE Transactions on Industrial Electronics*, vol. 55, pp. 4109–4126, December 2008.

[5] A. Stefani, A. Bellini, and F. Filippetti, "Diagnosis of induction machines' rotor faults in time-varying conditions," *IEEE Transactions on Industrial Electronics*, vol. 56, pp. 4548–4556, November 2009.

[6] A. Bellini, A. Yazidi, F. Filippetti, C. Rossi, and G.A. Capolino, "High frequency resolution techniques for rotor fault detection of induction machines," *IEEE Transactions on Industrial Electronics*, vol. 55, pp. 4200–4209, December 2008.

[7] A. Khezzar, M.E. Oumaamar, M. Hadjami, M. Boucherma, and H. Razik, "Induction motor diagnosis using line neutral voltage signatures," *IEEE Transactions on Industrial Electronics*, vol. 56, pp. 4581–4591, November 2009.

[8] O. Poncelas, J.A. Rosero, J. Cusido, J.A. Ortega, and L. Romeral, "Motor fault detection using a Rogowski sensor without an integrator," *IEEE Transactions on Industrial Electronics*, vol. 56, pp. 4062–4070, October 2009.

[9] T.M. Wolbank, P. Nussbaumer, C. Hao, and P.E. Macheiner, "Non-invasive detection of rotor cage faults in inverter fed induction machines at no load and low speed," *IEEE International Symposium, SDEMPED '09*, pp. 1–7, August/September 2009.

[10] M. Benbouzid, M. Vieira, and C. Theys, "Induction motors' faults detection and localization using stator current advanced signal processing techniques," *IEEE Transactions on Power Electronics*, vol. 14, pp. 14–22, January 1999.
[11] B. Yazici and G.B. Kliman, "An adaptive statistical time-frequency method for detection of broken bars and bearing faults in motors using stator current," *IEEE Transactions on Industrial Application*, vol. 35, pp. 442–452, March 1999.
[12] B. Akin, S. Choi, U. Orguner, and H. Toliyat, "A simple real-time fault signature monitoring tool for motor drive embedded fault diagnosis systems," *IEEE Transactions on Industrial Electronics*, vol. 58, pp. 1990–2001, May 2011.
[13] S.H. Kia, H. Henao, and G. Capolino, "A high-resolution frequency estimation method for three-phase induction machine fault detection," *IEEE Transactions on Industrial Electronics*, vol. 54, no. 4, pp. 2305–2314, August 2007.
[14] A. Bellini, G. Franceschini, and C. Tassoni, "Monitoring of induction machines by maximum covariance method for frequency tracking," *IEEE Transactions on Industrial Application*, vol. 42, no. 1, pp. 69–78, January/February 2006.
[15] W. Zhou, T.G. Habetler, and R.G. Harley, "Incipient bearing fault detection via motor stator current noise cancellation using Wiener Filter," *IEEE Transactions on Industrial Application*, vol. 45, pp. 1309–1317, July/August 2009.
[16] S. Rajagopalan, T.G. Habetler, R.G. Harley, J.A. Restrepo, and J.M. Aller, "Non-stationary motor fault detection using recent quadratic time-frequency representations," *Industrial Application Conference*, vol. 5, pp. 2333–2339, October 2006.
[17] S.M.A. Cruz, H. A. Toliyat, and A.J.M. Cardoso, "DSP implementation of the multiple reference frames theory for the diagnosis of stator faults in a DTC induction motor drive," *IEEE Transactions on Energy Conversion*, vol. 20, no. 2, pp. 329–335, June 2005.
[18] M. Blodt, D. Bonacci, J. Regnier, M. Chabert, and J. Faucher, "On-line monitoring of mechanical faults in variable-speed induction motor drives using the Wigner distribution," *IEEE Transactions on Industrial Electronics*, vol. 55, no. 2, pp. 522–533, February 2008.
[19] S.M. Kay, *Fundamentals of Statistical Signal Processing: Estimation and Detection Theory*, Englewood Cliffs, NJ: Prentice-Hall, 1993.
[20] C.-M. Ong, *Dynamic Simulation of Electric Machinery Using Matlab/Simulink*, Upper Saddle River, NJ: Prentice Hall, 1998.
[21] B. Akin, H. Toliyat, U. Orguner, and M. Rayner, "Phase Sensitive Detection of Motor Fault Signatures in the Presence of Noise," *IEEE Transactions on Industrial Electronics*, vol. 55, pp. 2539–2550, June 2008.
[22] G.B. Kliman, R.A. Koegl, J. Stein, R.D. Endicott, and M.W. Madden, "Noninvasive detection of broken rotor bars in operating induction motors," *IEEE Transactions on Energy Conversions*, vol. 3, pp. 873–879, December 1988.
[23] R. Schoen, T. Habetler, F. Kamran, and R. Bartfield, "Motor bearing damage detection using stator current monitoring," *IEEE Transactions on Industry Applications*, vol. 31, no. 6, pp. 1274–1279, November/December 1995.
[24] M.E.H. Benbouzid, "A review of induction motors signature analysis as a medium for faults detection," *IEEE Transactions on Industrial Electronics*, vol. 47, no. 5, pp. 984–993, October 2000.
[25] S. Nandi, M. Bharadwaj, and H.A. Toliyat, "Performance analysis of a three-phase induction motor under mixed eccentricity condition," *IEEE Transactions on Energy Conversions*, vol. 17, pp. 392–399, September 2002.

10

Electric Implementation of Fault Diagnosis in Hybrid Vehicles Based on Reference Frame Theory

Bilal Akin, Ph.D.
Texas Instruments

10.1 Introduction

The integrity of the electric motor in work and passenger vehicles can best be maintained by monitoring its condition frequently on-board the vehicle. This chapter presents a signal-processing-based fault diagnosis scheme for on-board fault diagnosis of rotor asymmetry at start-up and idle mode [9]. Regular rotor asymmetry tests are done when the motor is running at certain speeds under certain loads with stationary current signal assumption. It is quite challenging to obtain these regular test conditions for long enough time during daily vehicle operations. In addition, automobile vibrations cause a nonuniform air-gap motor operation, which directly affects the inductances of electric motors and results in a quite noisy current spectrum. Therefore, in examining the condition of an electric motor integrated to a hybrid electric vehicle (HEV), regular rotor fault detection methods become impractical. The presented method overcomes the aforementioned problems simply by testing the rotor asymmetry at zero speed. This test can be achieved and repeated during start-up and idle modes. The method can be implemented at no cost, basically using the readily available electric motor inverter sensors and microprocessor unit.

10.2 On-Board Fault Diagnosis (OBD) for Hybrid Electric Vehicles (HEVs)

It is very important for any vehicle to monitor its vital equipment continuously. Therefore, nowadays almost all vehicles are equipped with an on-board diagnostic (OBD) system [1]. This system has been used for warnings and monitoring critical failures in the vehicle, such as ignition, battery, oil and gasoline level, engine, and brakes. If a problem or malfunction is detected, the OBD system sets off a malfunction indicator light (MIL), readily visible to the vehicle operator on the dashboard, to inform the driver that a problem existed. When illuminated, it displays a universally recognizable symbol or a similar phrase for each failure. OBD is a valuable tool that assists in the service and repair of vehicles by providing a simple, quick, and effective way to pinpoint problems by retrieving vital automobile diagnostics from the OBD systems [2].

According to the U.S. Code of Federal Regulations (CFR), all light-duty vehicles, light-duty trucks, and complete heavy-duty vehicles weighing 14,000 pounds GVWR (gross vehicle weight rating) or less (including medium duty passenger vehicles [MDPVs]) must be equipped with an OBD system capable of monitoring all emission-related power train systems or components during the applicable useful life of the vehicle. A vehicle shall not be equipped with more than one general-purpose malfunction indicator light for emission-related problems; separate specific purpose warning lights (e.g., brake system, fasten seat belt, oil pressure, etc.) are permitted [2]. Although CFR's requirements for OBD are mainly related to environmental protection purposes, safety issues in vehicles should also be considered by using the OBD system.

The Code of Federal Regulations does not state any diagnostics requirements of electric machines in HEVs. Besides the battery, which is a vital electrical component in HEVs, monitoring the conditions of an electric machine is very critical in case of any failures such as bearing, rotor, and stator faults as shown in Figure 10.1. By diagnosing the electric machine faults as early as possible, one can prolong the lifetime of the electric machine in HEVs by performing maintenance before a catastrophic failure occurs. Therefore, emerging HEV systems require onboard fault diagnosis as shown in Figure 10.2, both to support critical functions of the control system and to provide cost effective maintenance.

A catastrophic failure in an electric machine might result in dangerous situations during driving, especially on the highway. Unless frequently monitored, an incipient fault in the machine can be propagated until it totally falls apart. Therefore, an accident afterward might become inevitable. Once the fault diagnostic system makes any kind of severe electric motor fault decision, the traction of the vehicle can totally be taken over by the combustion engine in order to prevent permanent damages and total loss of the electric motor. Basically, this solution is applicable if HEVs are designed based on parallel or parallel and series architectures. However, in series configurations, the internal combustion engine (ICE) is directly connected to the electric motor [3]. Therefore, in series architectures the solution is limited to electric faults and has partial use for mechanical faults such as bearing fault.

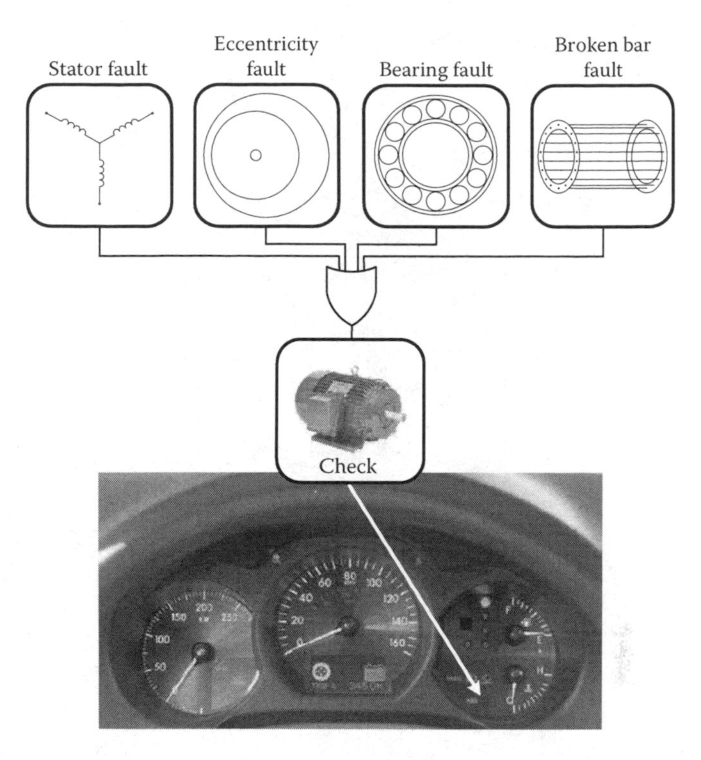

FIGURE 10.1
Motor fault can be displayed in the hybrid electric vehicle instrument cluster (Lexus GS 450h).

Mechanical vibration of the vehicle degrades the fault diagnosis of the electric motor integrated to the HEV. The vibration causes nonuniform air-gap operation, and therefore the machine inductance oscillates. Because of this oscillation, the line current becomes noisy and the noise floor of the current spectrum becomes higher. This noise in the current spectrum generated by mechanical vibration degrades the fault signature analysis results. Therefore, one of the best alternatives is condition monitoring at zero speed. Either idle modes or start-up might provide long enough time to process the current data and report the condition of the electric motor. On the other hand, the vibrating nature of the vehicle makes use of other vibration-sensitive sensors, such as the accelerometer, impractical due to the excessive noise at sensor output. Unlike accelerometers, flux sensors, and so forth that are mounted on the electric motor for fault diagnosis, the current sensors are located inside the motor drive unit, which is far away from the main source of vibration. Moreover, the cost of the current sensor is relatively low when compared to the other sensors. Thus, one of the best alternative combinations is employing current sensors at zero speed where mechanical vibration effect is minimum.

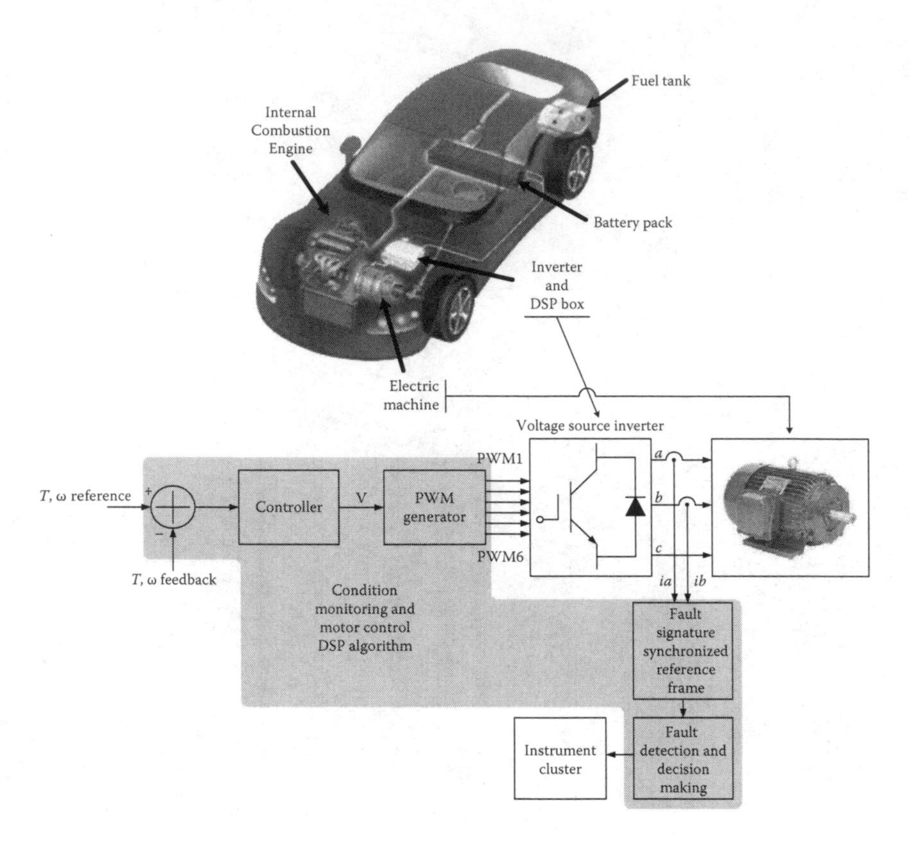

FIGURE 10.2
Drive-embedded fault diagnosis scheme integrated to HEV. (From IEEE Power Electronics Society Newsletter, vol. 19, no. 1, pp. 1, First Quarter 2007. With permission.)

10.3 Drive Cycle Analysis for OBD

Drive cycle is typically used by independent emissions testing laboratories to validate hybrid electric vehicle economy and emissions. The U.S. Environmental Protection Agency (EPA) city cycle is the first 1300s of the Federal Test Procedure, FTP75, regulated cycle shown in Figure 10.3. Table 10.1 shows the most common drive cycles and their statistics respective of their geographical regions [5]. Other than highway mode, traffic flow is uneven, with very frequent stop–go events and long idle times as shown in Table 10.1. This is why city cycles have low average speed compared to similar performance on the U.S. highway cycle. Because of the high percentage of stop time as shown in Table 10.1, the presented OBD algorithm can often be run to monitor the motor condition where the

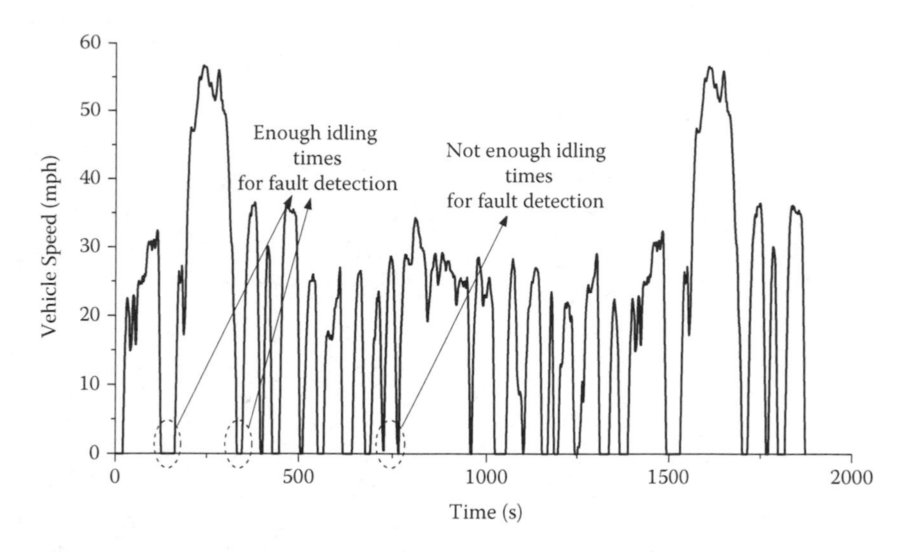

FIGURE 10.3
US FTP75 city drive-cycle and fault detection points during idling. (From Sambandan, S., Nathan, A., Single-Technology-Based Statistical Calibration for High-Performance Active-Matrix Organic LED Displays, Journal of Display Technology, Vol. 3, Issue 3, DOI: 10.1109/JDT.2007.900914, 2007, pp. 284 – 294. With permission.)

mechanical vibration is minimal. Since the current spectrum analysis is mostly based on transformed signal averaging, the transient state fault signature analysis has a high degradation potential. As shown in Figure 10.3, the drive cycle is dominated by transients where the motor current has nonstationary characteristics. Thus, instead of continuous condition monitoring, fault detection can be limited to start-up and idle modes in order to enhance the reliability of fault decision warning. Idle stop functionality, as shown in urban drive cycle in Figure 10.3, is the primary means by which there is fuel consumption reduction, and turns out to be a safe strategy for electric motor fault diagnostics. Every time the vehicle stops at a stop sign or traffic lights, or stop-and-go heavy traffic, the fault monitoring

TABLE 10.1

Standard Drive Cycles and Statistics

Region	Cycle	Time Idling (%)	Average Speed (kph)
Asia-Pacific	10-15 mode	32.4	22.7
Europe	NEDC	27.3	32.2
North America-city	EPA-city	19.2	34
North America-highway	EPA-highway	0.7	77.6
North America-US06	EPA	7.5	77.2
Industry	Real World	20.6	51

algorithm is run. If the stop (zero speed) time is not enough to finalize a fault decision, which is typically a few seconds, then the diagnostic result is neglected and resumed.

10.4 Rotor Asymmetry Detection at Zero Speed

Broken rotor bars in an induction motor rotor cause field asymmetry, which results in special sidebands at frequencies $f_s(1 \pm 2s)$ [6–7]. In real-time applications, a number of challenges must be considered to detect these sidebands. For example, because these signatures directly depend on the slip, the rotor speed should be measured precisely. Otherwise, without accurate enough speed information, it is not possible to distinguish broken rotor bar sidebands from the fundamental component. One alternative solution might be to eliminate a fundamental component using a notch filter for line-driven motors. However, this solution is not applicable to variable speed drive systems due to dynamically changing stator frequency. Furthermore, the notch filter might cause sideband suppression unless sufficient loading is not provided during the tests. Another alternative is to estimate the rotor speed during the operation, which brings extra computational burden. However, at low speed range most of the speed estimation algorithms cannot provide precise information. Therefore, high-speed operation must be guaranteed in order to obtain high precision speed values for sensorless broken rotor bar detection.

Next, the motor should be loaded at certain torque values smoothly in order to raise these sidebands and separate them from the fundamental component in the current spectrum. Smooth and proper loading might not be available for various applications to test the rotor asymmetry. The method detects broken rotor bars in real time without employing speed sensor and loading systems. Some other external hardware employed in previous works [6–7], such as the data acquisition systems and analog filters, is also eliminated.

The test is implemented at zero speed; therefore there is no need for speed measurement or speed estimation. The rotor is locked mechanically or electrically using direct current braking. Because the injected signal to test the rotor asymmetry is below 10% of rated voltage values, the generated torque during the test is negligible. Thus, the broken rotor bar test can be implemented without an additional featured loading system. The $f_s(1+2s)$ term is caused by torque vibrations [8] and cause electromechanical chain interaction between the rotor and the stator that result in many asymmetrical signatures in the spectrum. Therefore, low frequency range is dominated by consecutive asymmetry signatures that sophisticate fault analysis as shown in Figure 10.4a.

Since the slip is maximum when the rotor is stationary, the existing consecutive signatures are quite far away from each other, as shown in Figure 10.4b. Indeed, at zero speed these terms vanish theoretically and the spectrum is

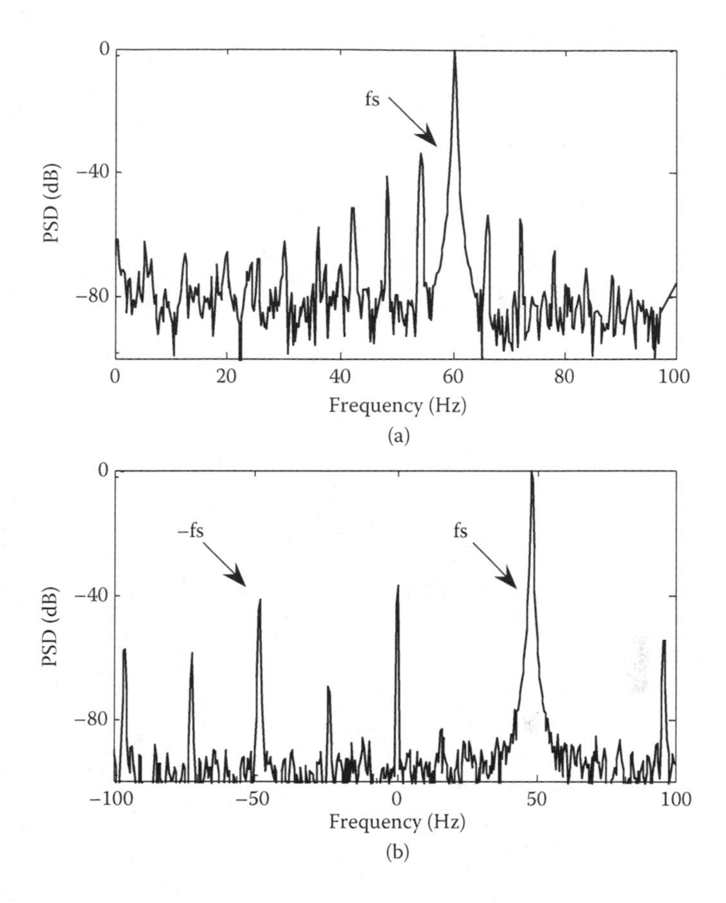

FIGURE 10.4
Current spectrum of broken rotor bar motor (a) regular test and (b) zero-speed test. (From Sambandan, S., Nathan, A., Single-Technology-Based Statistical Calibration for High-Performance Active-Matrix Organic LED Displays, Journal of Display Technology, Vol. 3, Issue 3, DOI: 10.1109/JDT.2007.900914, 2007, pp. 284 – 294. With permission.)

quite clean when compared to the current spectrum at rated speed and rated torque operation.

The method presented here focuses on $f_s(1 - 2s)$ term. At zero speed $s = 1$, thus the $f_s(1 - 2s)$ term is at frequency of $(-f_s)$. Single-phase reference frame analyses do not work for negative frequency; therefore three-phase current vectors are transformed to complex current space vector. Single-phase real or imaginary current analyses are insensitive to vector rotation direction; thus they find superposition of the left sideband and the fundamental component at the supply frequency (f_s). In order to compute the left sideband and the fundamental components separately, the current space vector fault relevant frequencies are experimentally examined.

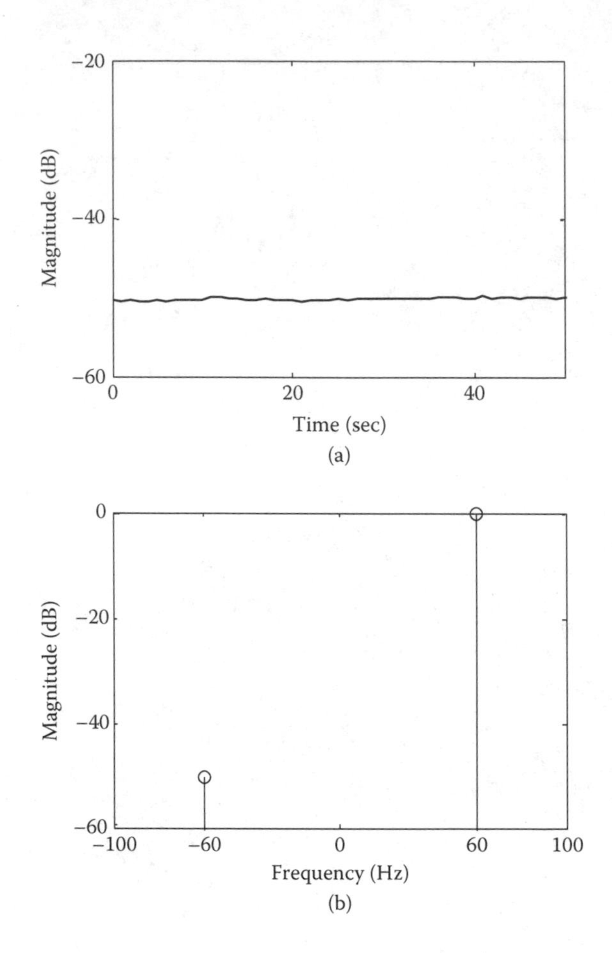

FIGURE 10.5
Normalized left sideband magnitude of a healthy motor obtained by the DSP in real time at standstill ($I = 9$ A, V/Hz = 1.0, $f = 48$ Hz): (a) time-frequency domain, (b) frequency domain.

In Figure 10.5 and Figure 10.6, the real-time fault signature tracking result is given when the motor is injected with low voltage at standstill near the rated current. Both the fault signature magnitude at ($-f_s$) and the fundamental component at (f_s) are computed simultaneously and separately in real time by the digital signal processor (DSP) in order to obtain the normalized fault signature magnitude.

In order to verify the method, a number of experiments are implemented under various volts/hertz ratios, line currents, and frequencies. It is reported that if high enough current is supplied around or higher than the rated current, under all conditions the healthy and faulty motors can be easily distinguished using this method. In Figure 10.5, fault signature frequencies of a healthy motor are examined when volts/hertz is set to 1.0 at 48 Hz. In Figure 10.5a, DSP continuously computes and updates the normalized left

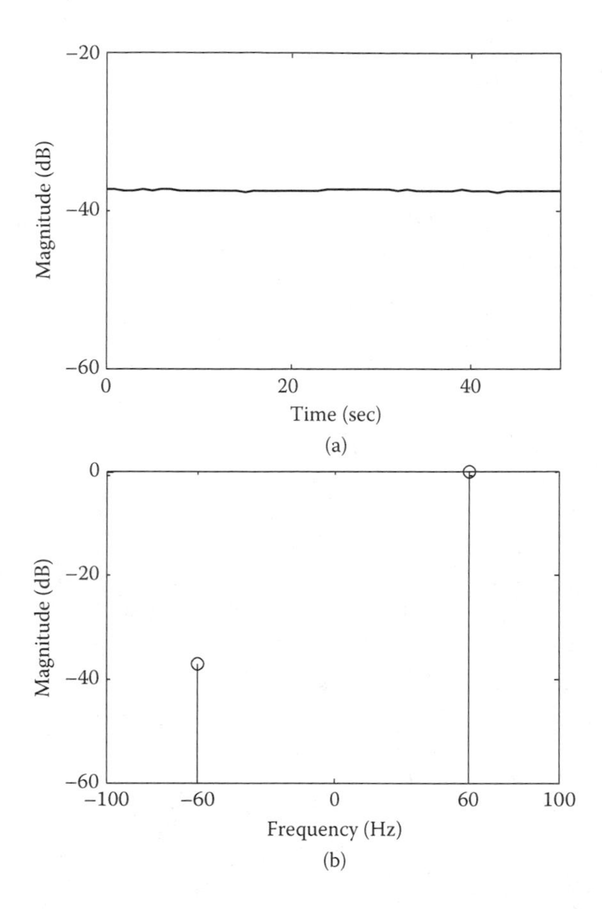

FIGURE 10.6
Normalized left sideband magnitude of a faulty motor obtained by the DSP in real time at standstill ($I = 9$ A, V/Hz = 1.0, $f = 48$ Hz): (a) time-frequency domain, (b) frequency domain.

sideband of the healthy motor current spectrum in each second. Figure 10.5b depicts an instant magnitude of left sideband component relative to fundamental in frequency domain.

A similar test is repeated for the motor that has less than 10 % broken rotor bars on the cage, and the results are shown in Figure 10.6. When Figure 10.5 and Figure 10.6 are compared to each other, it is clearly seen that under the same conditions the left sideband is increased by 13 dB, which is high enough to distinguish healthy and faulty motors from each other.

In Figure 10.7, the normalized fault component magnitudes are given at various frequencies when the voltage-to-frequency (v/f) ratio is equal to 0.5. The comparative results are as promising as the regular (full-load, rated speed) broken rotor bar test. It is reported that the differences between the left sideband of healthy and faulty motor fault signatures are very close to

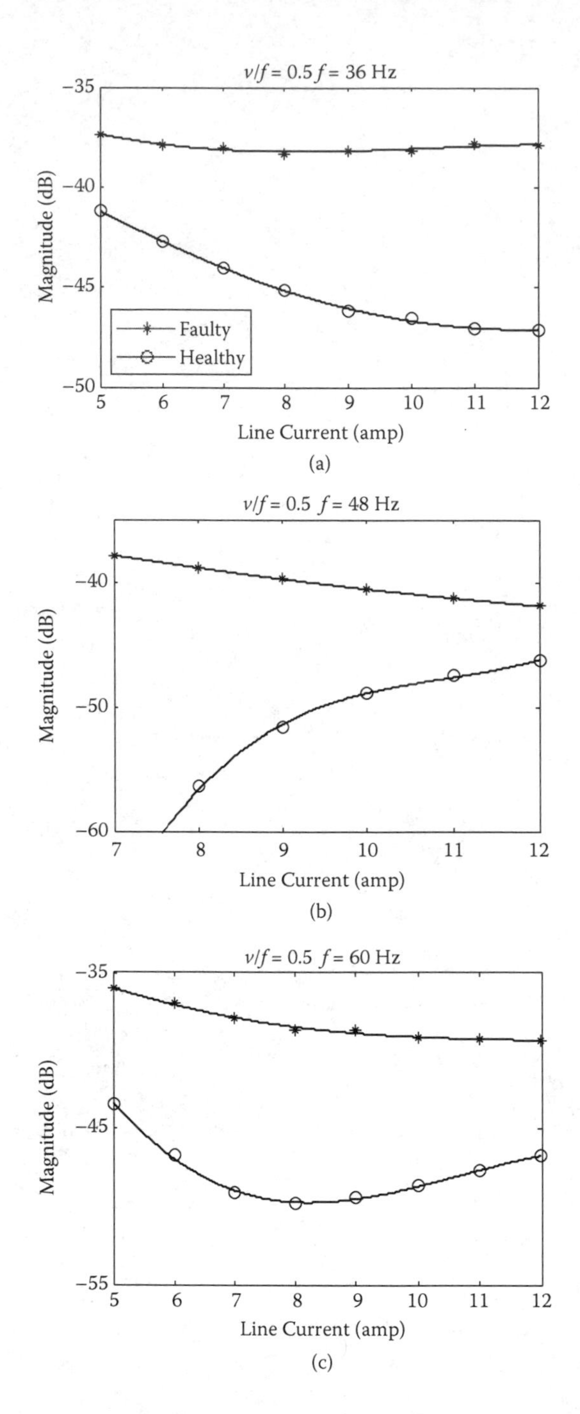

FIGURE 10.7
Normalized left sideband magnitude obtained by the DSP in real time versus line current (v/f = 0.5): (a) 36 Hz, (b) 48 Hz, (c) 60 Hz.

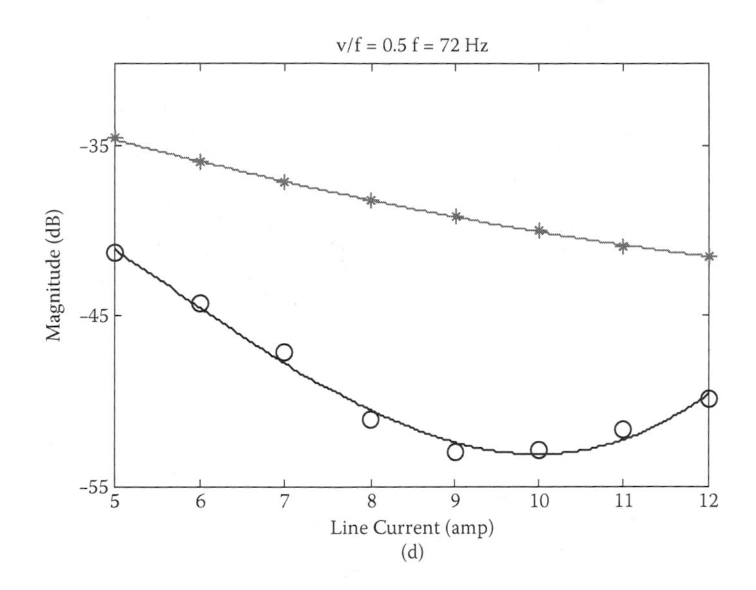

FIGURE 10.7 (continued)
Normalized left sideband magnitude obtained by the DSP in real time versus line current (v/f = 0.5): (a) 36 Hz, (b) 48 Hz, (c) 60 Hz.

regular test results. Therefore, the same results can be obtained at standstill without the need for any external hardware just before the motor start-up or at idle mode in a few seconds.

The same test is implemented when the v/f ratio is set to 1.0 to examine the effect of magnetizing current. It is noticed that the results are close to the ones obtained at two different v/f ratio tests as shown in Figure 10.8 and Figure 10.9. Thus, magnetizing current level has limited effect on the left sideband at standstill and can be ignored as a fault analysis parameter.

In conclusion, the condition monitoring and fault detection of electric motors in hybrid electric vehicles are quite vital for safety and cost-effective maintenance. This chapter therefore presents a simple on-line on-board fault diagnosis of induction motor for HEVs at start-up and idle (standstill) conditions based on reference frame theory. The major advantages of the method are very fast convergence time, no need for an additional sensor or hardware, robustness and reliablility, speed sensorless implementation, and zero-speed application, making it highly robust against the mechanical vibrations effects. It is experimentally shown that the method detects the rotor asymmetry fault signatures at start-up and idle mode (zero speed) and determines the severity of the fault. The solution can easily be extended to the other faults for complete motor monitoring.

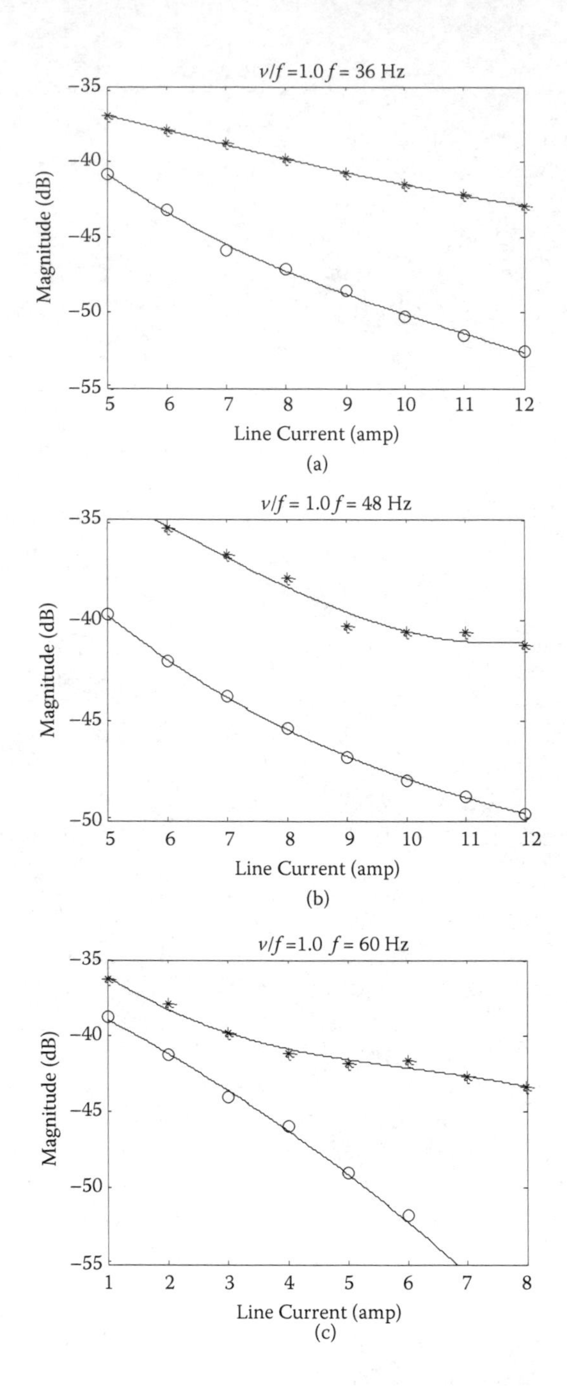

FIGURE 10.8
Normalized left sideband magnitude obtained by the DSP in real time versus line current (v/f = 1.0): (a) 36 Hz, (b) 48 Hz, (c) 60 Hz.

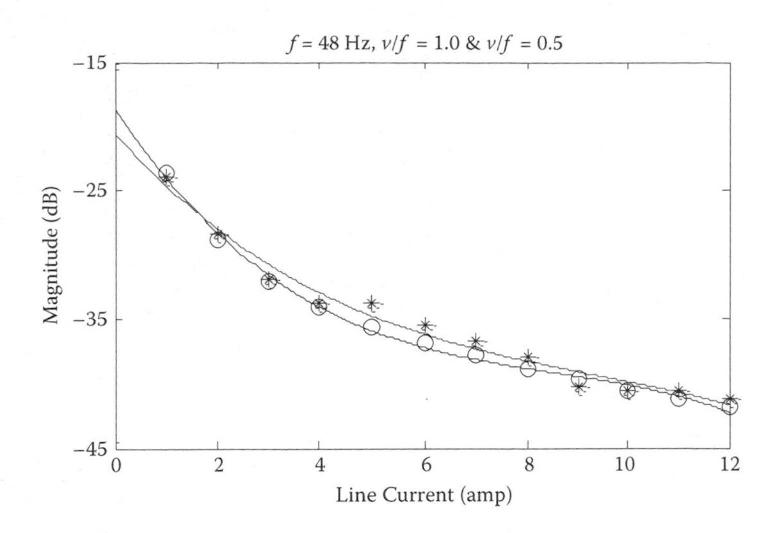

FIGURE 10.9
Normalized left sideband magnitude obtained by the DSP in real time versus line current (v/f = 1.0 and 0.5, f = 48 Hz).

References

[1] R. K. Jurgen, *On- and Off-Board Diagnostics*, Warrendale, PA: Society of Automotive Engineers, 2000.

[2] Office of the Federal Register (U.S.), Code of Federal Regulations, Title 40, Protection of Environment. Washington, DC: Office of the Federal Register, 2006.

[3] M. Ehsani, Y. Gao, S.E. Gay, and A. Emadi, *Modern Electric, Hybrid Electric, and Fuel Cell Vehicles: Fundamentals, Theory, and Design*, 1st ed., Boca Raton, FL: CRC Press, 2004.

[4] IEEE Power Electronics Society Newsletter, vol. 19, no. 1, pp. 1, First Quarter 2007.

[5] J. M. Miller, *Propulsion Systems for Hybrid Vehicles*, Stevenage, UK: Peter Peregrinus, 2004.

[6] C. Kral, F. Pirker, and G. Pascoli, "Detection of rotor faults in squirrel-cage induction machines at standstill for batch tests by means of the Vienna monitoring method," *IEEE Transactions on Industry Applications*, vol. 38, no. 3, pp. 618–624, May 2002.

[7] H. Henao, C. Demian, and G.A. Capolino, "Detection of induction machines rotor faults at standstill using signals injection," *IEEE Transactions on Industry Applications*, vol. 41, no. 6, pp. 1550–1559, November 2004.

[8] F. Filippetti, G. Franceschini, C. Tassoni, and P. Vas, "AI techniques in induction machines diagnosis including the speed ripple effect," *IEEE Transactions on Industry Applications*, vol. 34, pp. 98–108, 1998.

[9] B. Akin, S.B. Ozturk, H.A. Toliyat, and M. Rayner, "DSP-based sensorless electric motor fault-diagnosis tools for electric and hybrid electric vehicle power train applications," *IEEE Transactions on Vehicular Technology*, vol. 58, pp. 2679–688, January 2009.

11

Robust Signal Processing Techniques for the Implementation of Motor Current Signature Analysis Diagnosis Based on Digital Signal Processors

Seungdeog Choi, Ph.D.
Toshiba International

11.1 Introduction

The motor fault has commonly been categorized as electrical and mechanical fault of a motor in literature. In addition to the conventional fault, the failure of the fault diagnosis algorithm itself in the harsh industry application can be considered as another serious fault condition that fails to perform motor protection and increases the possibility of unwanted system failure. [8]

To implement a full fault-detection procedure using digital signal processors (DSPs) in industry, the applied techniques should not only correctly detect the fault signatures but also make reliable decisions. An effective algorithm should be able to take variations in fault signature amplitude, line current noise level, frequency offset, and phase offset into consideration in order to avoid missing or false detection alarms.

In practical applications, a small fault frequency offset between the expected and the existing fault signature frequency can be observed due to inaccurate speed feedback or estimation, slow response time of sensing devices. This offset can create an error in motor current signature analysis (MCSA) techniques used in industry. Even with tolerable speed feedback error in motor control, if the detection is performed within a short period, a small fault frequency offset can aggravate the overall capability of the speed-sensitive detection system. Therefore, it is unlikely to make a reliable decision regarding the fault status until the fault frequency offset is compensated accurately, which has commonly been neglected in many studies. The phase estimation of a fault signal requires another concern in fault diagnosis as it is commonly challenging to make a correct phase estimation of a small signal in a noisy channel. Also, if frequency errors exist, the phase estimation of a fault signal becomes practically and theoretically impossible. Noise level and its variations must also be considered in a diagnostic system design because the fault signatures are generally observed at a much smaller level than the noise energy level [2–4]. Due to the low signal-to-noise ratio (SNR), a robust fault detection method applied for plants in harsh industrial environments should accurately consider noise content and its variation.

Ignoring these ambiguities might result in erroneous fault indices in industrial applications. Furthermore, to come up with highly reliable fault indices based on fault references, the thresholds should be updated depending on the motor speed, torque, and control schemes, which will result in further complexity.

This chapter presents a comprehensive fault detection procedure that performs both the fault detection and decision-making stages taking nonideality into account and maintaining the complexity low enough for DSP-based, real-time implementation.

11.1.1 Coherent Detection

In signal processing, one of the well-known and most widely used detection methods is classified in two parts: coherent detection and noncoherent detection [5]. Coherent detection basically uses measured frequency and phase distortion of a signal, which is compensated in the subsequent stages of the fault detection. On the other hand, noncoherent detection is applied without knowing the phase information. Since precise measurement of inspected low amplitude fault signatures is a challenging task, noncoherent detection is a more practical tool for fault diagnosis applications. Indeed, once the necessary information is accurately provided, the coherent detection usually performs better than noncoherent detection as it utilizes more signal information, which increases the complexity [5]. The noncoherent detection yields more reliable detection under severely noisy conditions where inaccurate information is available as its performance is not dependent on the distortion factor.

A simplified coherent detection is presented by Akin et al. [2]. As shown in Figure 11.1, the fault amplitude and phase can be monitored using a phase

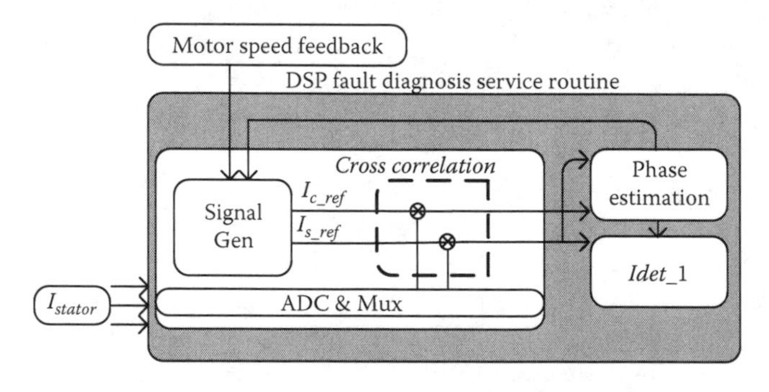

FIGURE 11.1
Coherent detection (phase-sensitive detection).

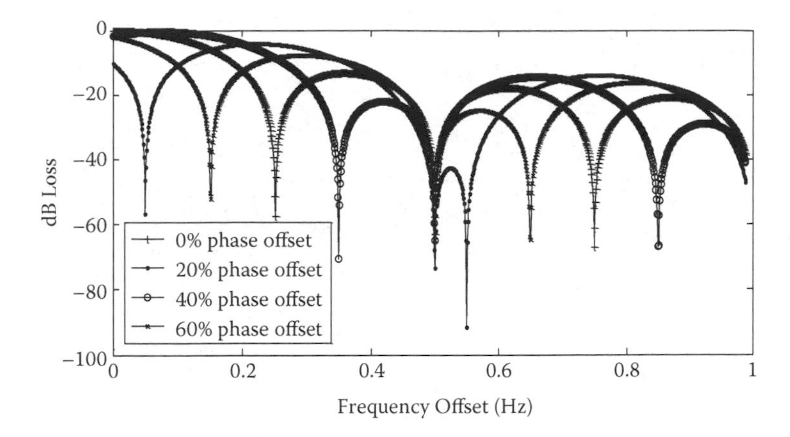

FIGURE 11.2
Fault signature detection loss versus frequency.

detection procedure. Compared to the techniques detailed by Bellini et al. [4] and Benbouzid et al. [7], phase-sensitive detection has reduced computational complexity and made it possible to perform a large amount of data processing using a low-cost DSP. However, the performance of coherent detection depends on the phase accuracy of the fault signature as depicted in Figure 11.2. Therefore, these kinds of techniques are applicable in conditions where phase ambiguities are negligible.

11.1.2 Noncoherent Detection (Phase Ambiguity Compensation)

The noncoherent detection is basically the amplitude detection procedure based on phase elimination and hence is inherently immune to the phase ambiguities of fault signatures. The elimination of the phase estimation stage reduces the computational burden of the noncoherent detection technique. Even though it shows lower performance than coherent detection at a steady state when the phase information is provided, it is well known that the detection method allows the user to obtain more reliable detection results under noisy or dynamic system conditions [5]. The block diagram of noncoherent detection is briefly given in Figure 11.3. So depending on the detection environment of fault SNR, the fault diagnosis method is to be determined between coherent and noncoherent detection.

11.1.3 Fault Frequency Offset Compensation

Rotor speed is one of the most critical variables that needs to be monitored continuously both for motor control and fault detection. The speed feedback is measured either by an encoder or estimated without speed sensors in the DSP code. Unlike the motor control, very precise speed

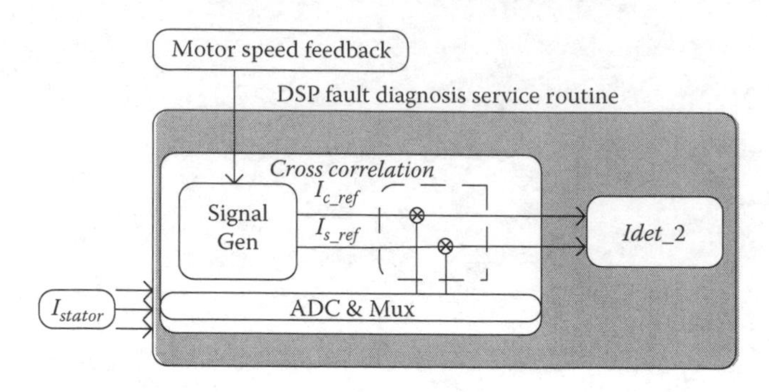

FIGURE 11.3
Noncoherent detection.

information is needed to identify the severity of the faults. However, in practical applications, a small mismatch between the speed feedback and the actual speed is commonly observed due to encoder resolution, inaccurate speed estimation algorithms, noise interferences, or slow response of sensors to unexpected transient condition, and so on. In addition to phase ambiguities, even a small amount of fault frequency offset yields erroneous fault detection results. Assuming the fault frequency offset $w_{offset} = w_{fault} - w_{ref} \neq 0$ and phase delay $\varphi_{ref} - \varphi_{fault} = \varphi_{offset}$, the cross-correlation output signal will be

$$\begin{aligned} I_{cross} &\approx K_1 \cos[\omega_{ref} n - \omega_{fault} n + \varphi_{ref} - \varphi_{fault}] + \varpi \\ &= K_1 \cos[\omega_{offset} n + \varphi_{offset}] + \varpi \end{aligned} \quad (11.1)$$

where ϖ is the motor current noise.

In Figure 11.2, the normalized decibel loss of coherent detection versus fault frequency offset is simulated using MATLAB where the phase offset percentage is defined between zero and 2π. It is clearly shown that fault frequency offset, phase offset, or a random combination of these two can truly suppress the fault signature, which typically has –40 to –80 dB amplitude.

However, great complexity will also be required if all of the expected offsets are monitored. Here, the current signal, $I_{cross}(n)$, is expected to provide high enough resolution for fault detection even if it is averaged in time or down sampled with noise elimination through averaging. The information of fault signature is expected to remain, as the low fault frequency offset will not be interfered with in the low pass filtering such as the averaging operation if it is appropriately designed based on the Nyqist theorem.

Applying an offset detection technique to an averaged signal with small samples will reduce the complexity. Maximum likelihood (ML) detection is used to estimate the sinusoid at offset frequency, which is the maximum of the periodogram [1] and is given by

$$\hat{f}_{ML} = \arg\max \left| \frac{1}{N} \sum_{n=1}^{N} x_n e^{j2\pi f n} \right| \tag{11.2}$$

where

$$x_n = \frac{1}{N_1} \sum_{k=1}^{N_1} I_{cross}[k + N_1(n-1)], \; n = 1, 2, \cdots N_2 \tag{11.3}$$

is the averaged signal, N_1 is the number of samples averaged, N_2 is assumed the physical DSP buffer size used for this purpose where the relation between parameters is as follows:

$$N = N_1 N_2 \tag{11.4}$$

The tracking bound without aliasing is given by

$$Track\_bound \le \frac{N_2}{2} \; Hz \tag{11.5}$$

The maximum bound comes from the Nyquist sampling theorem. If the offset ($w_{offset} = w_{fault} - w_{ref} \neq 0$) is assumed, the aliasing will not be observed, practically.

The computational complexity of ML detection in Equation (11.2) depends on N and the frequency range $f_{range} = f_{\max} - f_{\min}$. Since the ML algorithm application in this study has high complexity, it needs modification for real-time DSP applications. These parameters will be limited through the averaged (effectively down sampled) signal with reduced N and limited frequency range where the maximum fault frequency offset between the reference signal and the fault signal frequencies is assumed to be less than 1 Hz ($= f_{range} < 1Hz$) for simplicity. Since the frequency error is fundamentally caused by motor speed feedback error, it can be adaptively adjusted depending on the performance of a speed estimator in industry application. In this way, the ML estimator can effectively be utilized in a DSP for on-line fault diagnosis.

The frequency resolution of ML-based offset detection in Equation (11.2) is determined as follows:

$$f_{\Delta} = \frac{f_{\max} - f_{\min}}{N_{tri}} = \frac{f_{range}}{N_{tri}} \tag{11.6}$$

where N_{tri} is the number of applications of ML trials within f_{range}.

11.2 Decision-Making Scheme

Procedures in diagnostics commonly consist of several steps that are signature detection, decision making, and final feedback to the controller or human interface system. Application of a low-cost diagnostic system in the industry is limited by the capability to handle the detection and decision-making process simultaneously within the same microprocessor. Assuming the detection steps shown in previous sections, the applicability of the discussed system further depends on the complexity and reliability of the decision-making scheme.

11.2.1 Adaptive Threshold Design (Noise Ambiguity Compensation)

Reliability is one of the major challenges facing fault diagnostic systems because the decision should be made for a small fault signature in a highly noisy industrial environment. In fact, the detection algorithm applied at fault frequencies detects noise signatures even with healthy motors, the amplitudes of which are usually hard to be discriminated from small fault signatures. One of the practical design considerations of the threshold encountered is how the detected signature can be reliably decided as the existing fault signature. The diagnostic decision making based on the threshold trained to the motor line current noise variation can evaluate the reliability of detected signature in DSP applications.

Here, the threshold is derived using the statistical decision theory [1] with the hypothesis of H_0 and H_1 for decision tests, which are as follows:

$$H_0 : I_{stator} = \varpi, \quad H_1 : I_{stator} = I_{fault} + \varpi \quad with\ p(\varpi) - N(0, \sigma^2), \tag{11.7}$$

where H_0 is the hypothesis of having only noise without any faults; H_1 is the hypothesis of existing fault signature with amplitude I_{fault} in white Gaussian noise, ϖ, channel; and $N(0,\sigma^2)$ means zero mean noise with variance σ^2. I_{fault} is assumed reliably detected under the phase and frequency errors of a signal, which are the major errors in diagnostic signal processing. Therefore, the hypothesis in Equation (11.7) becomes possible by advantaging the techniques in previous sections independently derived from any control scheme of a motor assuming major error conditions.

A decision rule is made based on the optimal statistical test with a likelihood-ratio test (LRT) of the two distributions of this hypothesis, which is as follows:

$$\Phi(I_{stator}) = \frac{P(I_{stator} : H_1)}{P(I_{stator} : H_0)} > \gamma \tag{11.8}$$

where γ is the temporary threshold. With Gaussian distribution of noise, Equation (11.8) is derived as follows:

$$\Phi(I_{stator}) = \frac{\exp[-\frac{1}{2\sigma^2}]\sum_{n=1}^{N}(I_{stator} - I_{fault})^2}{\exp[-\frac{1}{2\sigma^2}]\sum_{n=1}^{N}(I_{stator})^2} > \gamma \tag{11.9}$$

$$\Rightarrow \qquad \frac{1}{N}\sum_{n=1}^{N} I_{stator} > \frac{\sigma^2}{NA}\ln\gamma + \frac{I_{fault}}{2} = \gamma' \tag{11.10}$$

where γ' is the threshold and N is the number of samples of current signal used for detection.

Let $T = \frac{1}{N}\sum_{n=1}^{N} I_{stator}$ in Equation (11.10). Then, the statistics of averaged stator current signal, T, is calculated as follows:

$$T - \begin{cases} N(0,\sigma^2/N) & Under\ H_0 \\ N(I_{fault},\sigma^2/N) & Under\ H_1 \end{cases} \tag{11.11}$$

The performance with threshold γ' applied to the averaged signal T with statistics shown in Equation (11.11) can be derived using the detection probability in Equation (11.12) and false alarm event probability in Equation (11.13), which are as follows:

$$P_D = \Pr\{T > \gamma'; H_1\} = Q\left(\left(\gamma' - I_{fault}\right)/\sqrt{\sigma^2/N}\right) \tag{11.12}$$

$$P_{FA} = \Pr\{T > \gamma'; H_0\} = Q\left(\gamma'/\sqrt{\sigma^2/N}\right) \tag{11.13}$$

where Q is the Q-function, which is detailed in a later section.

With range of allowable error (false alarm), P_{FA}, a threshold is calculated from Equation (11.12) and Equation (11.13):

$$\gamma' = \sqrt{\sigma^2/N}\, Q^{-1}(P_{FA}) \tag{11.14}$$

The threshold provides a reliable decision-making tool for small signature detection in a noisy channel. Signature-based fault diagnosis performed with reliably detected signatures through the threshold will lead to more accurate condition monitoring while discriminating its results from random noise interference signatures. From Equation (11.14), the threshold is dependent on the number of samples and the noise variance estimated. These are independently determined from the motor operating point parameters (i.e., the fundamental stator current level, torque, rotor speed, motor specifications). This is a desirable feature of the fault diagnosis algorithm applicable for general purposes.

It becomes possible since the complicated motor environments are generally reflected in line current noise, which is measured for threshold design. It implies that the diagnostic process is simplified without considering various reference estimations of different motor conditions, which will result in increased system complexity and prior knowledge of these variations.

The only unknown parameter in the threshold Equation (11.14) is the noise variance. The instantaneous line current noise is effectively measured for the threshold parameter using the method described later in this section.

11.2.2 Q-Function

Figure 11.4 (top) shows the probability distribution curve of noise and signature amplitude assuming an additive zero mean Gaussian noise channel. The area under each probability curve is one. By assuming an arbitrary decision threshold, γ_a, the probability distribution of decision-making errors can be identified in the shaded area as type I error. The reliability of small signal detection mainly depends on how the type II error (false detection) is suppressed. The Q-function is used to measure the error probability of false detection, which is the right side of the shaded area in Figure 11.4 (top).

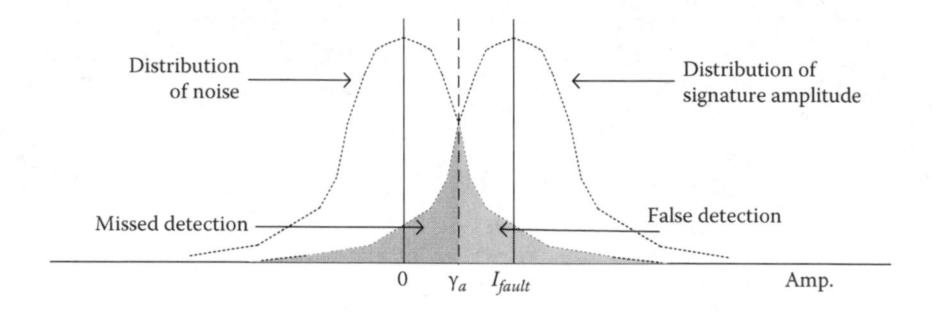

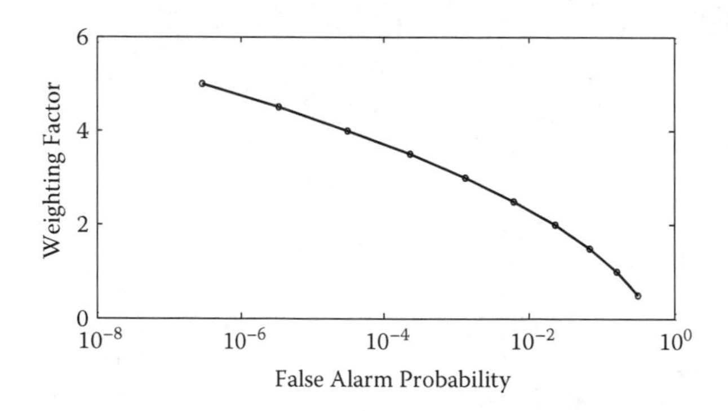

FIGURE 11.4
(Top) Probability distribution of diagnostic decision errors. (Bottom) Weighting factor versus false alarm probability.

The Q-function is defined as follows:

$$Q(z) = \int_z^{\infty} \frac{1}{2\pi} e^{-g^2/2}\, dg, \quad z = (\gamma_a - I_{fault})/\sqrt{\sigma^2/N} \tag{11.15}$$

The term $Q^{-1}(P_{FA})$ in Equation (11.14) is effectively a weighting factor. With a greater weighting factor, the threshold in Equation (11.14) is to be increased and the false detection rate is decreased; this relationship is computed through Equation (11.15) and plotted in Figure 11.4 (bottom). Once the allowable false error rate P_{FA} is determined and, hence, the weighting factor, diagnostic decision making can be performed with a constant false alarm probability independent from the random noise conditions of the line current signal. This is because the threshold in Equation (11.14) is adaptively determined based on instantaneous noise condition and decouples the effect of noise on decision-making performance.

The optimization of threshold level and parameter depends on the diagnostic requirement of a specific system. An ideal threshold simultaneously minimizes false detection and missing detection probability. In small signal detections, minimizing the false alarm is commonly of more concern. Based on assumed noise conditions and allowed error probability, threshold parameters can be adaptively designed and optimized for a target system.

11.2.3 Noise Estimation

Noise variance can be estimated via the mean squared error (MSE) criterion. The MSE estimation is performed assuming infinite estimation time and zero mean noises from uniformly distributed signal distortion. Since harmonic signals are approximately periodic in the stationary operation of a motor and averaged to zero, noise content remains as follows:

$$E_N = \frac{1}{N}\left[\sum_h \sum_{k=1}^{N} \left(e^{j(w_{harmonic(h)}k + \varphi_{harmonics(h)})}\right) + \sum_{k=1}^{N}(\varpi_k)\right] \approx \frac{1}{N}\sum_{k=1}^{N} \varpi_k \tag{11.16}$$

Let $\hat{\mu} = \frac{1}{N}\sum_{k=1}^{N} \varpi_k$ be an unbiased estimator of $\varpi \mu = E(\hat{\mu})$ whose mean is $E(\hat{\mu}) = 0$. From MSE criterion, noise statistics are derived as follows:

$$mse(\mu) = E[(\hat{\mu} - \mu)^2] = E[((\hat{\mu} - E(\hat{\mu})) + (E(\hat{\mu}) - \mu))^2]$$

$$= Var(\hat{\mu}) + (\mu - E(\hat{\mu}))^2 = Var(\hat{\mu}) = \sigma^2/N \tag{11.17}$$

$$\Rightarrow \sigma^2 \approx N Var(E_N) \tag{11.18}$$

where $\mu - E(\hat{\mu}) = 0$. Noise variance is derived from Equation (11.18).

11.3 Simulation and Experimental Result

11.3.1 Modeled MATLAB Simulation Result

A typical stator current is modeled with fault conditions of a broken rotor bar. The distorted current signal is established assuming –15 dB noise, and 11% total harmonic distortion (THD) with 5th and 7th harmonics. The broken rotor bar signature of –40dB amplitude is inserted based on the fault equation assuming slip $s = 0.016$ pu where the excitation frequency is 60 Hz. The simulation is performed with the modeled signal in Figure 11.5. (All the experiments/simulations are performed assuming steady-state operation of a motor.) In the simulation, P_{FA}, frequency tracking range and the available buffers of a DSP are assumed the same as shown in Table 11.1. The signal with 50K samples is utilized for each simulation result. The fundamental signal is assumed filtered in the simulation.

The frequency tracked amplitudes and the threshold measured are shown simultaneously in Figure 11.5 with offsets varying from 0 to 1 Hz. In the figure, negative (–) frequency values are simply replicas of positive (+) offset results for convenience since noncoherent detection cannot discriminate polarity of frequency. Zero frequency is the point where the tracking scheme is not applied. In the figure, it is shown that the fault frequency offset inserted is accurately tracked at the frequency of the maximum normalized amplitude in all trials. One can also determine that the maximum points are above threshold while signals are below each threshold at the

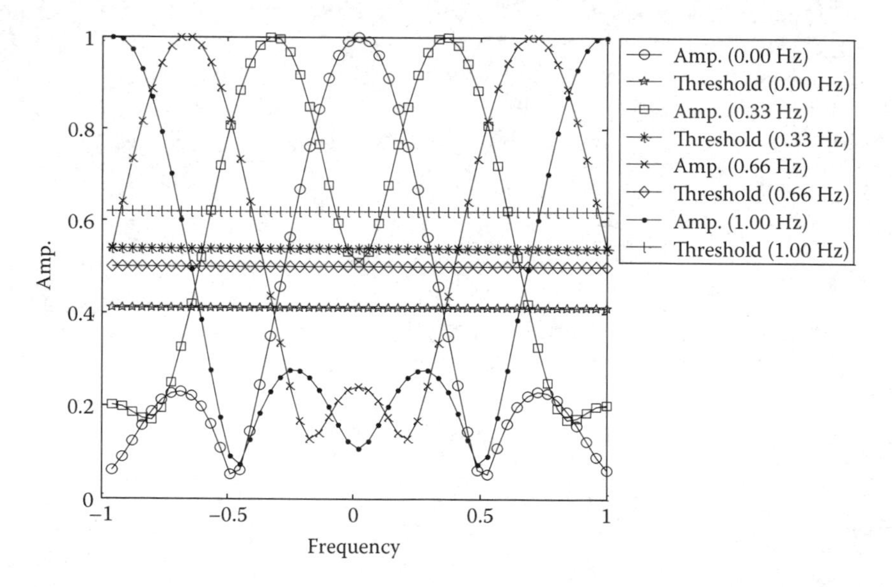

FIGURE 11.5
Frequency tracking with possible offsets (resolution: 0.04 Hz).

TABLE 11.1

Experiment Environment

Sampling Hz	25 kHz
Data acquisition board	NI-DAQmx
Motor	3 hp/4-pole IM
DSP board	eZ DSP 320F2812
Frequency tracking range	1 Hz
P_{FA}	0.00097
Buffer size (N_2)	500

point without frequency tracking. The assumed tracking resolution 0.04 Hz shows sufficient performance to discriminate maximum points.

11.3.2 Off-Line Experiments

The experiments are run utilizing line current data obtained by a 1.25 MS/s, 12-bit resolution data acquisition system, which is set to produce a 25 KHz sampling frequency. The 3-hp induction motors are loaded by the direct current DC generator, which is assumed open-loop controlled in all experiments. The acquired off-line data are processed through MATLAB.

In Figure 11.6, the stator current from the data acquisition card is shown with eccentricity (Figure 11.6a) and mixed fault signatures and unknown

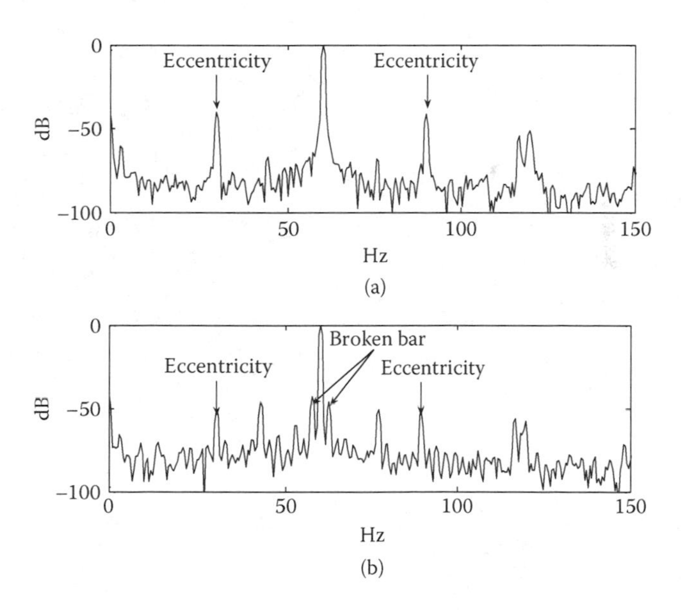

FIGURE 11.6
Stator current spectrum: (a) eccentricity signature at 20% torque, (b) mixed signature with broken rotor bar fault at 100% torque (supply frequency: 60 Hz).

TABLE 11.2

Decision Making 10 Seconds, Supply Frequency: 60 Hz, 20% Torque

DF	DA	DF_T	DA_T	TH
1	–40.16 dB	1	–40.16 dB	–49.78 dB

Note: DF, detection flag; DA, detected amplitude; TH, threshold. Subscript T is the result obtained through the frequency tracking operation.

signatures simultaneously through fast Fourier transform (FFT) analysis (Figure 11.6b). Instantaneous fault frequencies are measured based on the fault equation. The motor is designed with mixed fault conditions for performing the experiments in a practical environment.

In Table 11.2 and Table 11.3, DF is the detection flag, DA is the detected amplitude, and TH is the threshold. The definition with subscript T is the result obtained through the frequency tracking operation. All amplitudes are shown in decibels.

From an FFT spectrum analyzer, the eccentricity signature monitored is –41.2 dB at 20% torque. It tends to decrease in the high torque range and around –55.45 dB at 40%~100% torques. For the broken rotor bar signature, it is –45.7 dB at 50% torque. Unlike the eccentricity, the broken rotor bar signatures increase with load and –41.8 dB at 100% torque. These results are taken to evaluate the accuracy of detection in the off-line fault diagnosis.

11.3.2.1 Off-Line Results for Eccentricity

Correlations shown in Figure 11.3 are performed between motor current signal and reference fault signal which reference signal is generated based on motor speed-dependent fault characteristic frequency. Figure 11.7 shows the averaged correlation output (Figure 11.7a) and the frequency tracking result (Figure 11.7b). The averaged signal in Figure 11.7a is rounded due to the applied Hanning window to prevent the effects of spectral leakages in diagnostic signal processing. In Figure 11.7b, the maximum occurs at zero frequency, implying there is negligible fault frequency offset. The threshold is well placed to decide eccentricity fault. It is further confirmed in Table 11.2. The detected eccentricity signature is determined correctly in both trials of

TABLE 11.3

Decision Making 10 Seconds, Supply Frequency: 60 Hz, 100% Torque

DF	DA	DF_T	DA_T	TH
0	–49.8 dB	1	–41.32 dB	–42.30 dB

Note: DF, detection flag; DA, detected amplitude; TH, threshold. Subscript T is the result obtained through the frequency tracking operation.

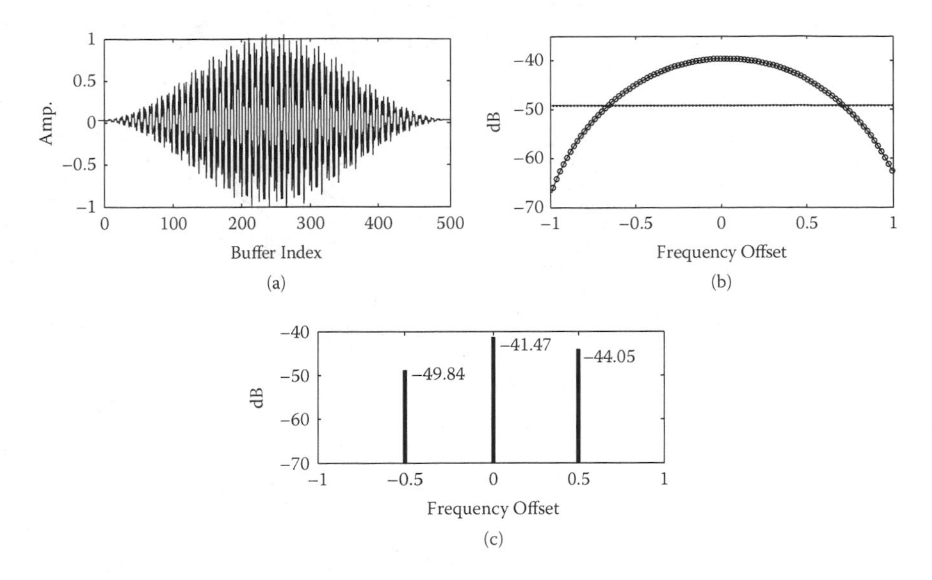

FIGURE 11.7
Frequency tracking for eccentricity fault: (a) averaged signal, (b) frequency tracking and decision making (resolution: 0.02 Hz), (c) coherent detection without strategy for fault frequency offset compensation.

frequency tracking and without tracking, with about 1.04 dB error from expected –41.2 dB obtained from the spectrum analyzer.

Figure 11.7c shows detection through the PSD scheme in Figure 11.1. The PSD is one of the algorithms utilizing optimal property of matched filtering, which has been adopted as a high-performance, low-cost fault diagnosis scheme. With the no-fault frequency offset (0 Hz) condition, the performance of the PSD is confirmed by the precise detection close to expected –41.2dB as shown in Figure 11.7c. With a potential frequency error at +0.5 Hz or –0.5 Hz, the analysis shows the loss of amplitude as expected in Figure 11.2. In the tracking scheme in Figure 11.7b, those frequency errors can be tracked and detections are compensated for reliable fault diagnosis. Because the schemes are optimized for precise detection in specific frequencies, serious loss of optimality occurs when the frequency/phase information has offsets as shown in Figure 11.7c. To be adopted in industry, robust performance under error conditions is to be maintained.

11.3.2.2 Off-Line Results for Broken Rotor Bar

In Figure 11.8a, the averaged signal is shown with dominant signal around 1.5 Hz. It is the fundamental stator current signal monitored at about 1.5 Hz away from the broken rotor bar signature (out of tracking range in Table 11.1). Although the technique is effective in small fault frequency offset tracking, it is inferred if the fundamental signal is within the tracking range. The range

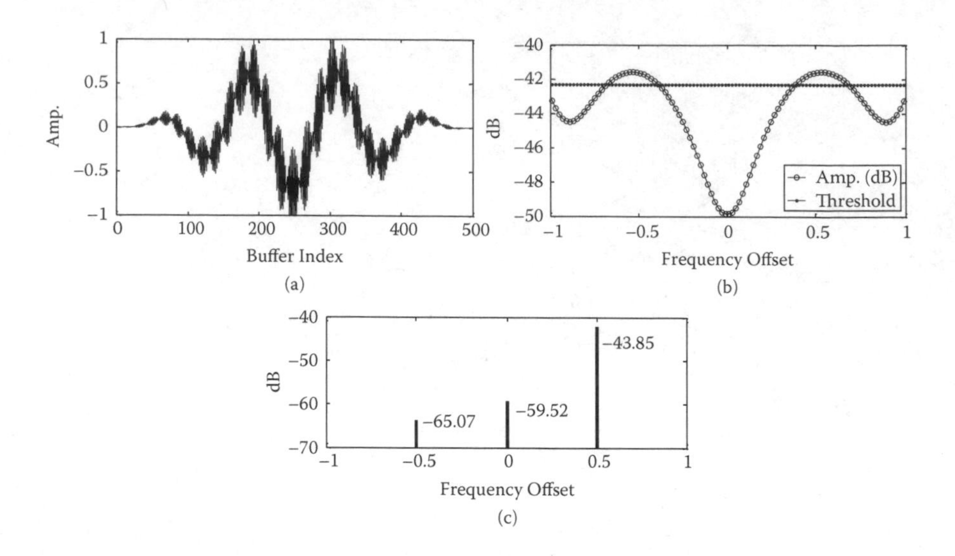

FIGURE 11.8
Frequency tracking for broken rotor bar fault: (a) averaged signal, (b) frequency tracking and decision making (resolution: 0.02 Hz), and (c) coherent detection without strategy for fault frequency offset correction.

need to be smaller than the difference between the supply and the expected fault frequencies.

In Figure 11.8b, the fault frequency offset is identified at maximum point with 0.46 Hz. In Table 11.3, the fault is determined correctly only after frequency tracking and detected amplitude is boosted from –49.8 to –41.32 dB. The accuracy of the ML tracking algorithm can be confirmed from the amplitude monitored through the spectrum analyzer, which is –41.8 dB and yields only 0.47 dB error from the tracked result.

Figure 11.8c shows detections through one of the optimal schemes, PSD, to compare the performance with the algorithm in Figure 11.8b under error conditions. Unlike the zero offset condition in Figure 11.7, the frequency/phase offsets are completely ambiguous in Figure 11.8. Figure 11.8c shows the serious performance degradation of amplitude loss due to frequency/phase ambiguity. Every detection at 0 Hz, –0.5 Hz, and 0.5 Hz shows unreliable values. Meanwhile, through the use of phase error-immunized detection and frequency tracking in Figure11.8b, the detection performance becomes close to optimal and robustness of detection is maintained under error conditions.

11.3.3 On-Line Experimental Results

The induction motor is fed by the inverter. The voltage-to-frequency (v/f) motor control and on-line fault diagnosis service routine are simultaneously implemented on a 32-bit fixed-point, 12-bit ADC, 150-MHz DSP of TMS320F2812.

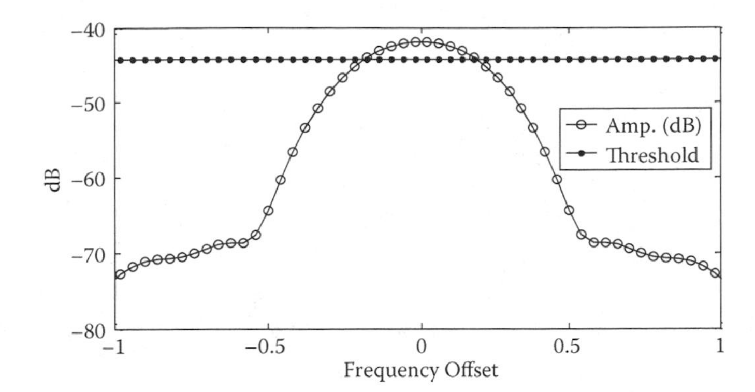

FIGURE 11.9
Frequency tracking for eccentricity signature with 20% torque at 10 seconds (supply frequency: 48.3 Hz, resolution: 0.04 Hz).

In Figure 11.9 and Figure 11.10, the zero frequency is the fault signature frequency measured by the DSP from the fault equation. In Figure 11.9, the DSP measures the fault signature frequency correctly showing a maximum at zero frequency, that is, −40.2 dB. In Figure 11.10, 0.24 Hz fault frequency offset between the expected and the existing fault signature frequency is monitored for the broken rotor bar signature.

The changes in detected amplitude and thresholds in time are shown in Figure 11.11 and Figure 11.12. In both figures, the detected signature hardly varies after 2 seconds. The threshold measured is unstable initially and becomes stabilized after about 8 seconds. After becoming stabilized, it tends to decrease since one of the threshold parameters, effective noise variance,

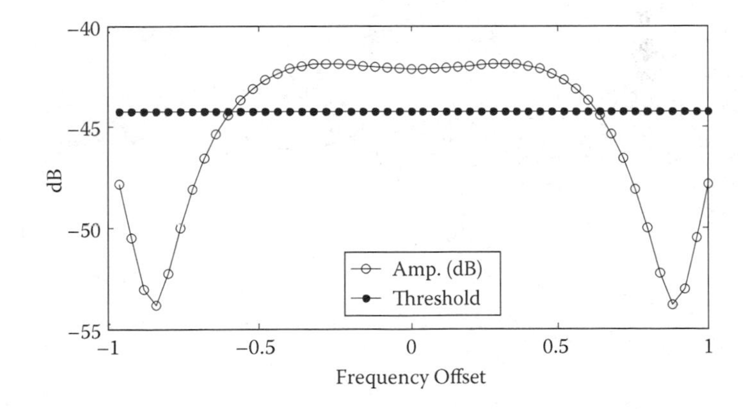

FIGURE 11.10
Frequency tracking for broken rotor bar signature (left side band) with 100% torque at 10 seconds (supply frequency: 48.3 Hz, resolution: 0.04 Hz).

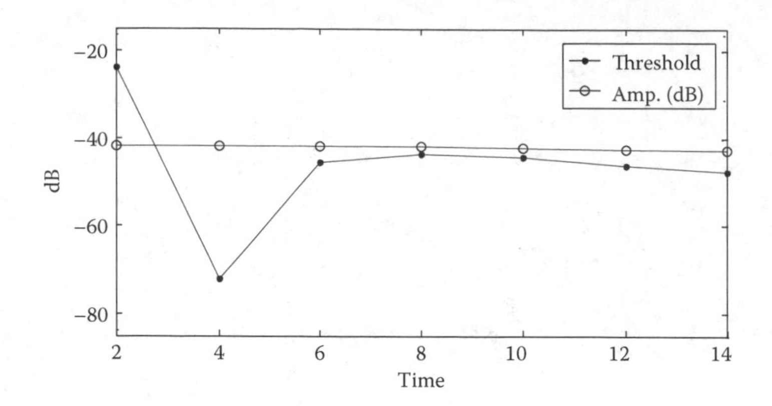

FIGURE 11.11
Detectability variation with time for eccentricity with 20% torque (supply frequency: 48.3 Hz).

σ^2/N , decreases as the number of samples used increases, which confirms careful derivation in Equation (11.14).

The latency time of about 10 seconds in fault diagnosis is assumed to be acceptable because condition monitoring is performed in a relatively long period of time, especially with a mechanical type of fault such as broken rotor bar or eccentricity.

In on-line experiments, the threshold applied is designed to keep false detection errors strictly within 0.097% as shown in Table 11.1. That is why the signatures are usually detected close to threshold within 5~10 dB. The thresholds can be further decreased to detect small signatures by reducing the weighting factor in Equation (11.14). This can be done based on the relation shown in Figure 11.4 (bottom) from the trade-off of detection performance.

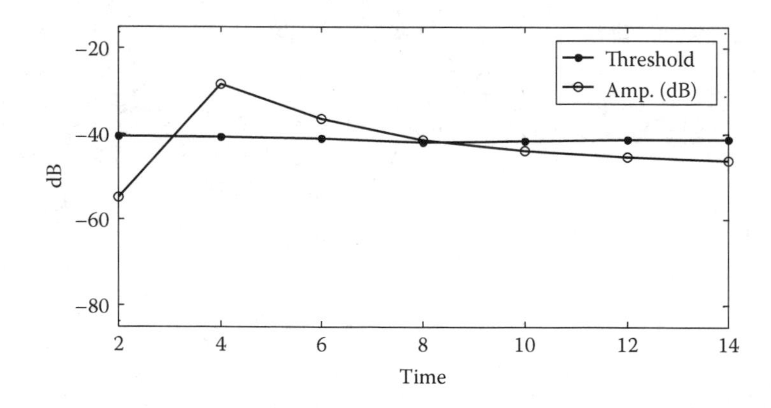

FIGURE 11.12
Detectability variation with time for broken rotor bar with 100% torque (supply frequency: 48.3 Hz).

The resolution of signature amplitude tracking can also be further improved by intentionally adding known frequency bias, which lets the detection achieved be more precise as the relatively high frequency signal can be identified in a relatively shorter time period.

The fault detection and decision-making capability of the robust fault diagnosis algorithm are demonstrated in this chapter by mathematical verifications and off-line/on-line experiments. It is observed that ambiguities such as the fault signature frequency mismatch, the phase of the fault vector, and changes in the noise level of fault signatures can be efficiently handled using a simple algorithm capable of frequency tracking, phase eliminating detection, and adaptive threshold.

References

[1] S.M. Kay, *Fundamentals of Statistical Signal Processing: Estimation and Detection Theory*, Englewood Cliffs, NJ: Prentice-Hall, 1993.

[2] B. Akin, H. Toliyat, U. Orguner, and M. Rayner, "Phase sensitive detection of motor fault signatures in the presence of noise," *IEEE Transactions on Industrial Electronics*, vol. 55, pp. 2539–2550, June 2008.

[3] S.H. Kia, H. Henao, and G. Capolino, "A high-resolution frequency estimation method for three-phase induction machine fault detection," *IEEE Transactions on Industrial Electronics*, vol. 54, no. 4, August 2007.

[4] A. Bellini, G. Franceschini, and C. Tassoni, "Monitoring of induction machines by maximum covariance method for frequency tracking," *IEEE Transactions on Industrial Applications*, vol. 42, no. 1, pp. 69–78, January/February 2006.

[5] A.J. Viterbi, *Principles of Coherent Communication*, New York: McGraw-Hill, 1966.

[6] S.M.A. Cruz, H.A. Toliyat, and A.J.M. Cardoso, "DSP implementation of the multiple reference frames theory for the diagnosis of stator faults in a DTC induction motor drive," *IEEE Transactions on Energy Conversion*, vol. 20, no. 2, pp. 329–335, June 2005.

[7] M. Benbouzid, M. Vieira, and C. Theys, "Induction motors' faults detection and localization using stator current advanced signal processing techniques," *IEEE Transactions on* Power Electronics, vol. 14, pp. 14–22, January 1999.

[8] S. Choi, B. Akin, M. Rahimian, and H.A. Toliyat, "Implementation of a fault diagnosis algorithm for induction machines based on advanced digital signal processing techniques," *IEEE Transactions on Industrial Electronics*, vol. 58, no. 3, pp. 937–948, March 2011.

Index

Z